Historical & Cultural Astronomy

More information about this series at http://www.springer.com/series/15156

Johann Friedrich Julius Schmidt

The Moon

A Translation of *Der Mond*

Translated by Stephen Harvey

With the Collaboration of Nicolas Matsopoulos

Springer

Johann Friedrich Julius Schmidt (1825–1884)

Translated by
Stephen Harvey
Horsham, United Kingdom

English translation of the German work "Der Mond" by J.F. Julius Schmidt, published by Barth, Leipzig, 1856

ISSN 2509-310X ISSN 2509-3118 (electronic)
Historical & Cultural Astronomy
ISBN 978-3-030-37271-2 ISBN 978-3-030-37269-9 (eBook)
https://doi.org/10.1007/978-3-030-37269-9

This Springer imprint is published by the registered company Springer Nature Switzerland AG
The registered company address is: Gewerbestrasse 11, 6330 Cham, Switzerland

Introduction

Johann Friedrich Julius Schmidt was one of the most dedicated and tireless visual observers of the heavens and other natural phenomena, both as an amateur and professional.

Julius Schmidt's interest in astronomy started from an early age and he was a compulsive observer of just about everything in the sky, but his enduring interest was the Moon. As W.H. Pickering said, Schmidt "… perhaps devoted more of his life than any other man to the study of the Moon." (Cited in Dobbins and Sheehan, 2014). That interest was spurred by the work of fellow Germans, Johann Hieronymous Schröter (1745–1816; Cunningham, 2014) and the lunar mappers, Wilhelm Beer (1797–1850; Baum, 2014) and Johann Heinrich von Mädler (1794–1874; Joeveer, 2014). Beer and Mädler's (1834) *Mappa selenographica : totam lunae hemisphaeram visibilem complectens observationibus propriis: quatuor sectionibus constructa et delineate* was, at that time, considered to be the 'be-all and end-all' of lunar cartography, and ironically its perceived perfection led to a fading of interest in further observation of the Moon. Julius Schmidt was to reverse that trend. So who was Julius Schmidt?

Fig. 1 Schmidt in Athens. (Courtesy National Observatory of Athens (copy from Prof. Ostwald Thomas collection))

Julius Schmidt: A Brief Biography

Johann Friedrich Julius Schmidt was born in Eutin, the Grand Duchy of Oldenburg, in Germany, on 26 October 1825 (Dobbins and Sheehan, 2014). His parents were Carl Friedrich Schmidt, a glazier, and Maria Elizabeth Schmidt, née Quirling.

As a child Schmidt went to school in Hamburg, where he developed a keen interest in nature. When he was 14 years old, he accidentally encountered the world of the heavenly bodies. As he says in his short biography, included in his monumental work *Charte der Gebirge des Mondes nach eigenen Beobachtungen in den Jahren 1840-1874* (Schmidt, 1878: IV-VIII):

> *"In Autumn 1839, in my homeland Eutin, in an auction I have found Schröter's book (Selenotopographische Fragmente, Vol I Lilienthal 1791, Vol II Gottingen 1802) about the Moon. The impression of the shadowed mountains and craters was so intense and lasting, that determined the rest of my life. I was only 14 years old and although I had been involved in zoology and botany for quite a while and I knew about some astronomical phenomena, there could be no definitive decision. I decided only when I was able to see the Moon's surface through a telescope. This desire was fulfilled soon, because a small but good telescope made by my father, showed me the numerous Moon craters. After I put it on a street lamppost, I spotted the rays of the Tycho crater and made my first sketch. The study of Schröter's book and the continuation of making Moon sketches became my main occupation...because of which I neglected my lessons..."*

In July 1841 the Hamburg Gymnasium (where he studied) made a school visit to Altona Observatory, whereby Dr. Petersen fascinated young Schmidt with views of the craters of the Moon. Here, he also was able to study the map of Beer and Mädler (which he would later refer to as his 'working lunar catalogue').

Whilst living in Hamburg, he was a frequent visitor at Hamburg Observatory and was soon entrusted as a volunteer observer by Dr. Christian Karl Ludwig Rümker (1788–1862; Holland, 2014) who allowed him to use the instruments of the Observatory in the years 1842–1845. Schmidt published his first astronomical report in *Astronomische Nachrichten* in 1843, (Nr.468) detailing his observations of variable stars and the Sun in 1841 and 1842.

When Schmidt was just 20 years of age; Professor Johann Friedrich Benzenberg (1777–1846; Kokott, 2014) offered him an assistant's position at his private Observatory at Bilk[1], near Dusseldorf. During this period, he was tasked with observing meteors and naked eye objects, as well as searching for possible intra-Mercurial planets using small telescopes. However, he was unable to continue his lunar observations, particularly with the primary telescope, as Professor Benzenberg was afraid that the "… looks and polish …" (Laios, 1962: 20) would suffer if Schmidt were to use it. The Professor died the following year (1846) and Schmidt moved onto another observatory.

In 1846 Schmidt took up an Assistant's position at Bonn Observatory with Professor Friedrich Wilhelm August Argelander (1799–1875; Markkanen, 2014). At Bonn his work was to measure the visual magnitudes and the positions of the

[1] Bilk Observatory was destroyed during a WWII air raid. A telescope memorial now stands on the site.

stars laying in the V hour sector of the celestial sphere, for the completion of the most famous star catalogue named "*Bonner Durchmusterung Des Nördlichen Himmels*", or as it is generally known 'B.D', and its accompanying star map. That was an enormous task because it contains the visual magnitudes and the celestial coordinates of 325,037 stars to visual magnitude 9–10, of the northern hemisphere of the sky (declination zones +89 to –1 degrees). Rather impressively, this catalog is that it is still in use by the astronomers (Batten, 1991).

During his stay at Bonn (1845–1853) his duties of making routine observations prevented him from making significant observations of the Moon, but he did find time to process older observations and prepare many drawings of lunar features based upon his previous measurements. Furthermore, he was able to observe the Moon during two visits (April 1849 and May 1853) to the Berlin Royal Observatory, where the famous Gottfried Johann Galle (1812–1910) provided him access to the 9.6-in. refractor.[2]

In 1848 he made an important set of drawings of Saturn during the equinox (a selection is reprinted in Lardner's "*Handbook of Astronomy*" – although incorrectly attributed to M. Julius Schmidt).

Argelander recommended Schmidt to take up the Director's position of the private Observatory of Baron Eduard Ritter von Unkrechtsberg (1790–1870) in Olmütz (today Olomouc, Czech Republic), Moravia, in 1853, where he stayed for almost six years until 1858. There he could work, as he wished, without any restrictions, or obligation for routine observations.

Fig. 2 Young Schmidt at Olmütz. (From Etching at Akademie der Wissenschaften (Wien))

[2]This was the same instrument that was used by Johann Gottfried Galle (1812–1910), in 1846, to confirm the existence of the planet Neptune.

Schmidt's work led him to participate in 1854 with the Curator of the Natural History Museum in Bonn, Thomas Dickert (1801–1883), in the construction of a giant relief model (half-globe) of the Moon's visible hemisphere. At over 19ft in diameter (see Fig. 3) the model was described in his book *Das Relief* ... (Schmidt,1854b) and illustrated in a 1925 leaflet issued by the Field Museum of Chicago (Farrington, 1925). The vertical relief on this model was exaggerated by a factor of three, compared to what Schmidt felt the correct relief should be. Although attracting much interest at the time, the present-day whereabouts of this remarkable model is unknown.

Fig. 3 A cropped photograph of the 19 ft. diameter semi-hemispherical model of the Moon made by Thomas Dickert and Julius Schmidt in 1854. (From http://www.idaillinois.org/cdm/search/collection/fmnh2/page/33)

During March and April 1855 Schmidt visited both Naples and Rome, making use of the great refractor in Rome to carefully map selected areas of the Moon; in Naples he made careful measurements of the heights of lunar mountains and craters with the use of a micrometer attached to the telescope, resulting in the publication of his book *Der Mond*... (Schmidt, 1856b).

Furthermore, during all this period he managed to continue many of his other observational projects - which actually continued for all his life – regarding the heights of various lunar formations (Schmidt, 1854a), the Zodiacal Light (Schmidt, 1856a), meteor showers (Schmidt, 1852b), sunspots covering a whole 11 years solar cycle (Schmidt, 1857a) and the observation of eclipses, especially the total Solar Eclipse of July 28, 1851(Schmidt, 1852a). Besides these major publications, Schmidt was a regular contributor to Argelander's prestigious journal *Astronomische Nachrichten.*

But that was not all. Schmidt had a deep interest in Natural Science. Thus, he was observing and recording any kind of natural phenomena he could. He visited volcanoes in order to study them, such as Etna and Vesuvius in Italy, where he measured the heights of different sites whilst determining the performance of a newly developed aneroid barometer (Schmidt, 1856d).

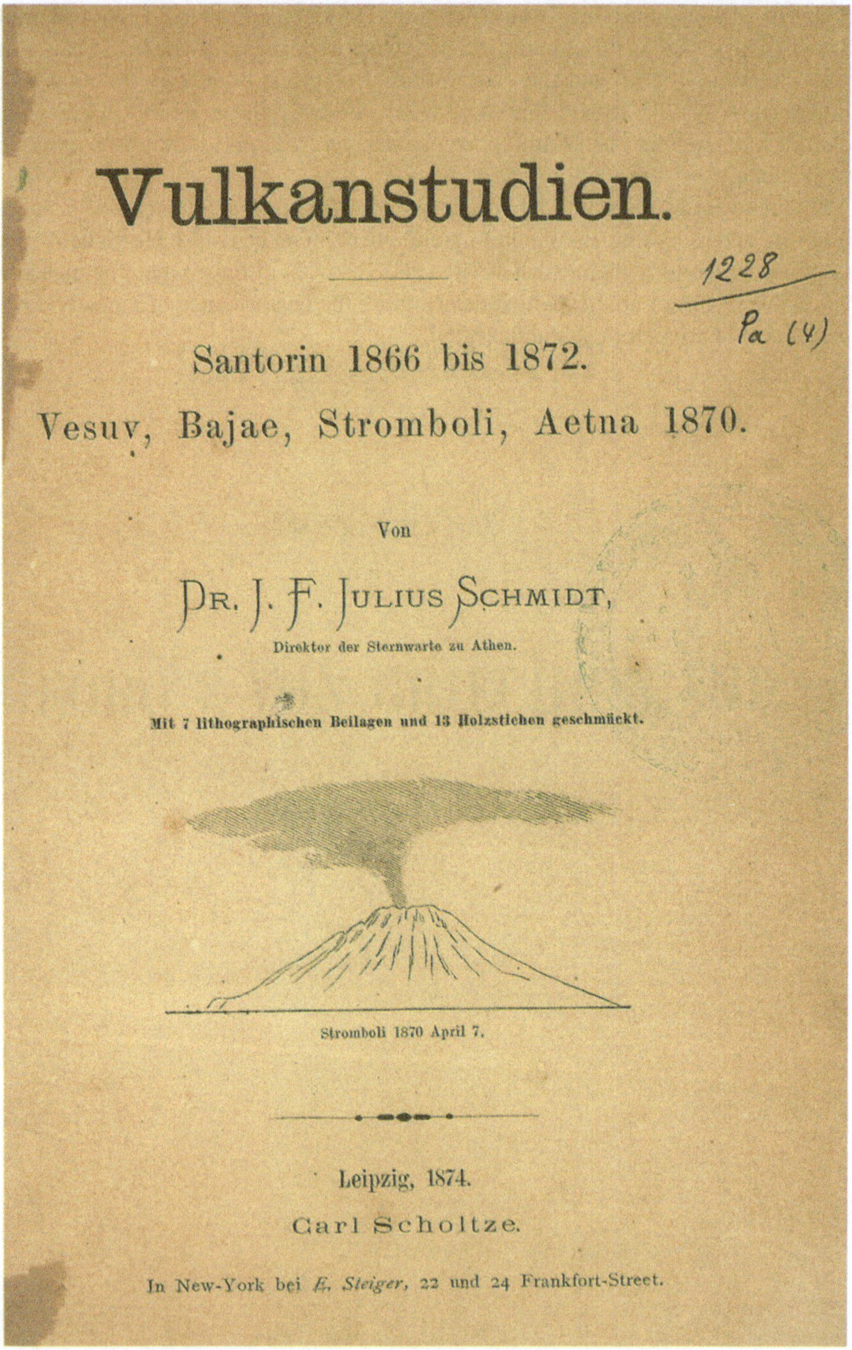

Fig. 4 Cover page from his book *Vulkanstudien* (Schmidt, 1874b). (From National Observatory of Athens)

He also recorded and studied earthquakes, resulting in two publications on the subject (Schmidt, 1857b; 1858b)

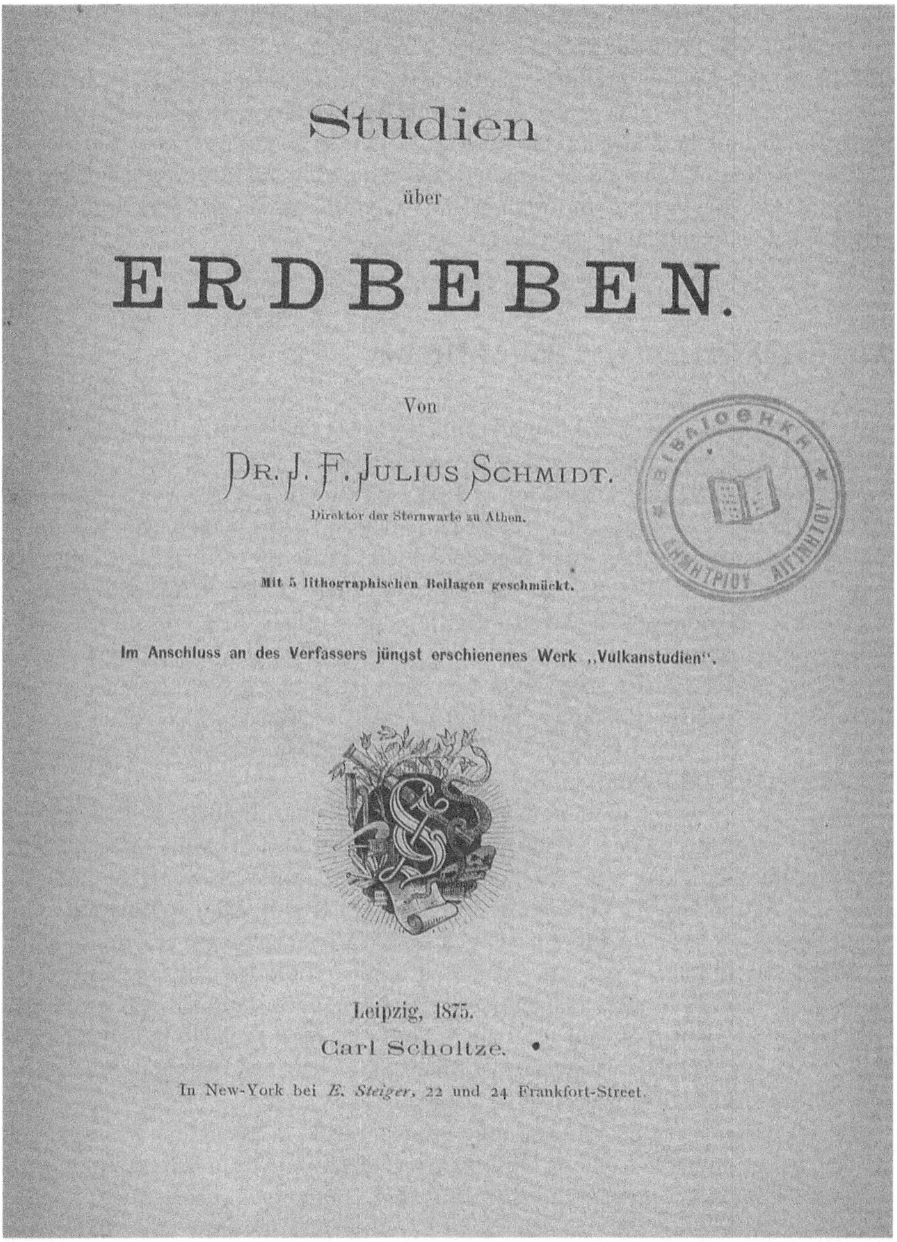

Studien

über

ERDBEBEN.

Von

Dr. J. F. Julius Schmidt.

Direktor der Sternwarte zu Athen.

Mit 5 lithographischen Beilagen geschmückt.

Im Anschluss an des Verfassers jüngst erschienenes Werk „Vulkanstudien".

Leipzig, 1875.

Carl Scholtze.

In New-York bei *E. Steiger*, 22 und 24 Frankfort-Street.

Fig. 5 Cover page from his book *Erdbeben* (Schmidt, 1875). (From National Observatory of Athens)

This rigorous activity provided him with a high reputation in the scientific community as a meticulous observer, although he had no formal training as a scientist.

In 1858 a very attractive proposal that could change his life, convinced him once again to relocate to a small country in the south-eastern part of Europe.

Baron Simon Sinas (1810–1876), a very wealthy merchant, banker and industrialist from Vienna, proposed Schmidt become the new Director of Athens Observatory in Greece. Sinas would provide Schmidt with a very respectable salary, from his own resources and absolute freedom to pursue his own research. Schmidt was never married and thus he had the freedom to make the decision for himself. Thus, he accepted the tempting proposal, which would secure him all that he wanted: a good salary for his living, freedom to pursue his scientific quests and above all a good astronomical site with more than 300 clear nights per year.

Athens Observatory: A Potted History

In 1830 Greece had achieved its independence from Ottoman-Turkish rule, after a revolution that began in 1821 and a struggle which lasted 9 years. London's Convention approved the second son of King Ludwig I of Bavaria prince Otto as the first modern King of Greece in 1832. Because Otto was very young, his government was initially run by a three-man regency council made up of Bavarian court officials. One of them was Anton von Prokesh-Osten (1795–1876), the Austrian Ambassador to Athens, who played a very important role in the establishment of the Athens Observatory.

Bavarians tried to organize the new-born state according to German standards. In their plans were the establishment of the University of Athens which began to operate in 1837. The first professor of Astronomy and Mathematics appointed was George Bouris (1790–1860).

Bouris was of Greek descent (actually from Ioannina - Epirus) but he was born in Vienna where his parents migrated to sometime before 1784 with numerous others from this particular area. He studied at Vienna University where he had as a teacher in Astronomy, the famous J.J. Littrow (1781–1840). After completion of his studies he worked as Director of the Greek school at Vienna for ten years until 1836.

Driven by an intense patriotic zeal, Bouris moved to Greece at the end of the same year in order to offer his services to the new-born state. He was hired as a translator at the Austrian Embassy where he met and be friended the ambassador Prokesh-Osten. When the University of Athens first opened, he was appointed as the first Professor of Mathematics and Astronomy.

He was a good mathematician and had experience in astronomy since he had worked as a volunteer at Vienna Observatory under Littrow. He had also published a paper on Biela's comet ("*Elliptische Bahnberechnung des Biela'schen Cometen mit Berücksichtigung sämmtlicher Bahn-Elemente und unmittelbarer Benützung der beobachteten Rectascensionen und Declinationen, aus sechs und neunzig*

Beobachtungen des Jahres 1832." Annalen der Kaiserlich-Königliche. Sternwarte in Wien.Wien, Part 14, p XXXVII-LIV).

It was Bouris who conceived the idea of establishing an astronomical observatory in Greece. In 1840, when he learned that Baron George Sinas (1783–1856), who was also the Greek Ambassador in Austria and the Chief Director of the National Bank of Austria, wished to donate a large amount of money to the Greek government in order to be used for the advancement of science at the University of Athens, he persuaded him to finance the establishment of an observatory. Obviously, Bouris and Sinas knew each other since both were eminent persons in the Greek community at Vienna. Furthermore, Sinas was from Epirus too, and actually from a small town called Moschololis (Now this area belongs to Albania). Prokesh-Osten embraced, supported and promoted the idea by all his means, especially to the King of Greece (Otto), mentioning that such an Institution would be very helpful in education, scientific research and navigation.

Fig. 6 Portrait of Baron George Sinas founder of the Athens Observatory. (From a portrait in the Victor Wimpffen (Vienna) collection)

Thus, the construction of the new observatory, on the top of the hill of Nymphs by the Acropolis, officially began on July 26 1842. Indeed, the cornerstone ceremony was timed to occur during a solar eclipse (embedding a fragmentary block used from the previous building that was already engraved "Hill of the Nymphs" in Greek). It was a formal ceremony attended by all the authorities, including the royal family and the King himself and practically most of the Athenians. Bouris, who had already been appointed the first Director of the Athens Observatory, addressed the crowd with an excellent speech, in which he referred to the high status of Astronomy in Ancient Greece, the contribution of the ancient Greek scholars and philosophers, and focused on the benefits of the existence of such an important institution for the country.

The building of the Observatory took 3 years to be completed under the auspices of the Danish architect Theofil Hansen (1813–1891). The result was a beautiful neo-classical cross-shaped building, with its four wings oriented toward the four cardinal directions and a central dome to house the main telescope.

Fig. 7 Original painting of the Athens Observatory by the architect who built it, Theofil Hansen. (From Akademie der Bildenden Kunste Wien)

Fig. 8 Portrait of the architect Theofil Hansen. (By Karl Rahl. Museum der Stadt, Wien)

Fig. 9 Theofil Hansen in Athens. (National Observatory of Athens)

Meanwhile, Bouris visited Vienna to purchase the first instruments for the Observatory. These included:

- A Plössl equatorial refractor of 6.2 in. aperture (158 mm, f/15) of dialytic design. Such a design employs a smaller concave lens, or combination of lenses, of high dispersive power, placed at a distance in the narrower part of the converging cone of rays, usually near the middle of the tube in order to correct the chromatic aberration of the objective lens.
- A Starke-Fraunhofer transit circle of 3.7 in. aperture.
- A Berthoud mean time clock.
- A Kessels sidereal time clock.
- Five small telescopes for comet hunting.
- Two Kapeller barometers.
- A set of other meteorological instruments.

In 1845 Bouris began his observations at the Observatory. He was making regular meteorological observations, meridian observations for the determination of the coordinates of the Observatory and the determination of time. He also observed the stars and the planets. His observations resulted in several scientific publications, mainly in *Astronomische Nachrichten* but the main corpus of his work remained unpublished and unfortunately is now lost. It is worth mentioning that at that time the Observatory had no other scientific personnel except its Director. Bouris worked completely alone.

Fig. 10 Portrait of George Bouris, first Director of Athens Observatory. (From National Observatory of Athens)

Bouris' efforts were undermined by another Professor of Mathematics at the University, Ioannis Papadakis who had allies at the ministry of Education. Obviously, Papadakis wanted the prestigious position of the Director for himself. Outrageous allegations were made against Bouris, such as: he published his scientific results in foreign languages and not in Greek; his work didn't support the navigation of the Greek merchant fleet; the students of the university had not gained from his research etc. The most absurd allegation was that he proposed to demolish buildings of the ancient city in order for the Observatory to have a better view of the southern part of the sky. Joseph Ashbrook in his book *The Astronomical Scrapbook* (1984) mentions this allegation, thereby sustaining the misinformation. The truth is that the southern horizon of the Observatory was, and still is, free of obstacles. The war against him was intense, causing frustration and depression for Bouris, and ultimately undermining his health, leading to him leaving Greece and returning to Vienna in April 1855, where he died five years later.

Bouris' Publications:

- *EllipticscheBahnberechnung des Biela'schenCometenmitBerücksichtigungsämmt licherBahn-Elemente und unmittelbarerBenützung der beobachtetenRectascensionen und Declinationen, aussechs und neunzigBeobachtungen des Jahres 1832*. Annalen der Kaiserlich-Königliche. Sternwarte in Wien. Wien, Part 14, pp. xxxvii-liv (1834)
- *Sur la variabilite du mouvementprope de Sirius en ascension droite*. – In: *Astronomische Abhandlungen als Ergänzungshefte zu den Astronomischen Nachrichten,* 119 (1849)
- *L'opposition de Neptune en 1848*. – In: *Astronomische Abhandlungen als Ergänzungshefte zu den Astronomischen Nachrichten,* 135 (1849)

- *Sur l'ascensiondroite de la Lune en 1847.* – In: *Astronomische Abhandlungen als Ergänzungshefte zu den Astronomischen Nachrichten,* 152 (1849)
- *Nachrichten von der Sternwarte Athens's nebst Beobachtungen der Irene ander selben. Astronomische Nachrichten,* 33(780),192–200 (1851)
- *Sur l'observatoired'Athenesetsur les extremes de temperature observes danscetteville. Archives des Sciences Physiques,* 24, 253–259 (1853)
- *Die Opposition des Mars imJahre 1849-1850 nach Beobachtungen der Sternwarte Athens's, nebst Bemerkungen über den Durchmesser des Mars. Astronomische Nachrichten,* 37(874), 153-188(1853)
- *Uber die Sirius - Tage. Astronomische Nachrichten,* 37(882), 311–316 (1853)

Ioannis Papadakis was appointed temporary Director of the Observatory, but he was not capable of any kind of scientific research. Thus, the Observatory was practically abandoned and fell into disrepair. Bouris, deeply concerned about the future of the Observatory, proposed to Simon Sinas to hire Schmidt as Director. Simon was the son of George Sinas, and his successor as benefactor to the Observatory after his father's death.

Fig. 11 Baron Simon Sinas. (Depicted in the book *Honpolgarok* (Barabas, 1866))

Schmidt at Athens

Julius Schmidt accepted Sinas' proposal and he was appointed the Director of Athens Observatory on 2 December 1858. Upon commencement, he immediately requested that repair and maintenance of the building and instruments be carried out. Baron Sinas agreed to cover the expense and the restoration took place in 1861.

Records show that the Meridian Circle was sent for repair to Vienna and equipped with four micrometers, (the equatorial refractor was also repaired again in the same city in 1874). The library also received many new important books, and several small instruments (mostly micrometers), were added to the Observatory's instrumentation.

During these changes, Schmidt began a series of initial astronomical and meteorological observations, focusing mainly upon observing meteors and variable stars. Afterwards, he set to work, with rigorous activity, to organise a regular observational service.

Schmidt, during his stay in Athens systematically studied:

1. Twilight
2. Sunspots
3. Meteors
4. Zodiacal light.
5. Photometric variable stars and others.
6. Positions and physical constitution of comets.
7. The rotation period of the major planets.
8. The colour of the stars.
9. The light, the appearance, the changes, the heights of the mountains and in general the description of the Moon.
10. The size of the diameters of the big planets.
11. The rings of Saturn.
12. The satellites of the big planets.
13. The positions and the physical constitution of nebulae.
14. Solar and lunar eclipses.
15. Use of the meridian circle and sextant to determine time.
16. Meteorology.
17. Hypsometric determination.
18. The Greek vegetation.
19. Seismology.

During the 25 years that he worked at the Athens Observatory, the clear skies allowed him to make thousands of naked eye observations of meteors as well as tens of thousands of observations of variable stars, discovering five periodic variables and two novae (T Coronae on 13 May 1866 and Nova Cygni on 24 November 1876). Most of his results were published in numerous research papers of the journal *Astronomische Nachrichten*. Schmidt also added a further 18 objects to the *New General Catalogue* (NGC): on 8 February 1861 he discovered the emission nebula complex NGC 6726, 6727 and 6729 in the constellation crater. Between 1845 and 1867 he determined the positions of 110 nebulae found by Herschel and Messier (see Schmidt, 1868). For many years Schmidt studied the planets and especially Mars and Jupiter, and he recorded their changing features in more than six hundred drawings. He observed the Great Comets of 1860 (C/1860 M1) and 1861 (C/1861 J1 (Tebbutt)), and one year later discovered the periodic comet, C/1862 N1 (Schmidt).

In order to gain a clear idea about his observing activity during this period of his life, we must note that of the 111 papers that appeared in *Astronomische Nachrichten*, 9 were about the planets, 35 about comets, 40 about variable stars, 6 about the Sun and solar eclipses, 4 about meteors and the rest presented mixed observational data regarding various objects. To underline the prolific amount of work carried out, we can see from the Table below a summary of observational reports presented to the Dean of Athens University for the years 1880 (Dean's Report … 1881: 90; Dean's Report … 1882: 71):

	1880	1881
Sunspot observations	359	357
Jupiter	230 drawings - measure of its rotational period	500 observations of Jupiter plus a few of Venus, Mercury and Mars
Moon	230 new measurements and drawings of various features. One lunar eclipse (December16, 1880)	708 new measurements and drawings of various features. One lunar eclipse (December 5, 1881)
Comets	5 comets, many positional and brightness observations	7 comets, many positional and brightness observations
Meteors	40 nights. 122 meteor paths	Few observations
Variable stars	Total magnitude estimations up to date 40336	476
Publications in *AN*	9	12

Furthermore, routine meteorological observations and star transits for the determination of time were carried out all year round. In his reports, Schmidt mentions the systematic assistance by the Professor of Astronomy at the University of Athens, Dimitrios Kokkides, especially in meridian observations, while Alexander Vourlis, who was his assistant, did most of the routine meteorological and sunspot observations at the Observatory.

Fig. 12 Wit Starke meridian circle used for the determination of time. (Courtesy of Theofanis Matsopoulos)

He also continued his observations of the Zodiacal Light in order to update his book *Das Zodiacallicht …* (Schmidt, 1856a).

Not content with studying just astronomy, Schmidt also published numerous books in the Natural Sciences. A pioneer of Greek seismology and with the help of volunteer observers he managed to record more than 3,000 earthquake descriptions, on the basis of a questionnaire. The results were published in his book "*Studien über Erdbeben*" (Schmidt, 1875), and in some monographs in Greek, such as "*On the 23rd of January 1867 Earthquake of the Kephalonia*" (Schmidt, 1867a).

He also studied the famous 1866 eruption of Santorini volcano from a Greek warship anchored off the island, and he published his study of this and of four other volcanoes, Bajae, Etna, Vesuvius, and Stromboli in "*Vulkanstudien – Santorin 1866 bis 1872. Vesuv, Bajae, Stromboli. Aetna 1870*" (Schmidt,1874b).

Fig. 13 Santorini (Thera) volcano eruption (drawing). (From Schmidt's book "*Vulkanstudien – Santorin 1866 bis 1872. Vesuv, Bajae, Stromboli, Aetna*", Leipzig (1874b))

Fig. 14 Santorini(Thera) volcano eruption (drawing). (From Schmidt's book "*Vulkanstudien – Santorin 1866 bis 1872. Vesuv, Bajae, Stromboli, Aetna*", Leipzig (1874b))

Schmidt travelled a lot in various places in Greece and out of its borders carrying out meteorological and geographical observations. He regularly sent these data to the Paris Observatory.

In the autumn of 1863, he accompanied the Austrian consul on his journey crossing the areas of Drinus and Axios, which was published under the title "*Reise durch die Gebiete des Drin und Vardar. Band I-II*", Wien (Hahn, 1867). The annex of the second volume of this work is his treatise titled "*Bemerkungenuber die Geographischen Ortsbestimmungen, Reise im Herbste 1863*" (Hahn, 1863).

Another interesting trip he made – with his architect friend Ernst Ziller–was to Burnarbaschi in Asia Minor, accompanying Hahn to his excavations searching for ancient Troy. The maps in Hahn's 1865 book "*Die Ausgrabungen vom Homerischen pergamos in zwei Sandschreibenan Georg Finlay*" were made by Schmidt.

Realising George Bouris' idea for producing regular publications of the work of the Observatory, he published two series of books in German, under the general title "*Publikationen der Sternwarte zu Athen*" (*Publications of the Athens Observatory*) financed by Simon Sinas.

The first book of these series does not contain astronomical research but it is about the natural geography of Greece -"*Beitragezur Physikalischen Geography von Griechenland*",Athens (Schmidt, 1861), while the second volume of this book (1864), is devoted to meteorological, hypsometric, and physical geography work in general. In volume I of Series I, Schmidt published observations of comets and mainly a long study of Comet Donati - "*Astronomische Beobachtungen über Cometen.*" (Schmidt, 1863), with several impressive drawings.

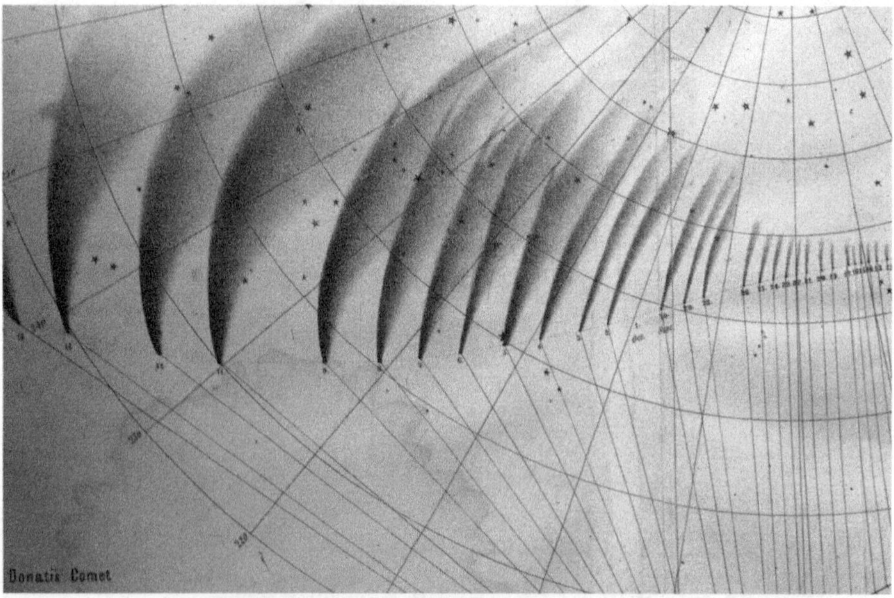

Fig. 15 The appearance and the path of Donati's comet on the sky by J. Schmidt. (From Schmidt's 1863 book *Astronomische Beobachtungen über Cometen*)

Subsequently, he published a book containing his observations of meteors and meteor showers, *"Astronomische Beobachtungen über Meteorbahnen und deren Ausgangspunkte"* (Schmidt, 1869) and the book *"Description physique d'Attique. Meteorologie et Phenomenologie"*(Schmidt, 1884).

He also wrote a large number of research papers and observational reports on various scientific subjects, which were submitted to the following publications:

- *Peterman's Midilungen*
- *Reports of the Academy of Vienna Science, Revisited by Heiss*
- *Proceedings of the Paris Academy of Sciences*
- *Reports of the Geological Institute of Austria*
- *Bulletins of the Royal Geographical Society of London*

However, his major accomplishment, which made him famous worldwide, was the compilation and drawing of a large topographical map of the Moon, including accompanying descriptive text: *Charte der Gebirge des Mondes – nach eigenen beobachtungen in den Jahren 1840-1874"* (Schmidt, 1874a). It was the result of 34 years of lunar observations. This exquisite map was eventually paid for by the Prussian Academy of Sciences, which financed its printing and publishing in 1878, along with the accompanying volume of explanatory notes. It is mainly for this work, which shows 32,856 distinct features compared to 7735 plotted by Beer and Mädler and 7178 by Lohrmann, that Schmidt is known. Schmidt also catalogued 348 rilles, compared to 71 for Beer and Mädler. He probably made more measurements and estimates of peak heights (~3000) and crater depths (~1000) than all other selenographers combined, and many of his measurements have yet to be replaced by modern ones.

His initial map was started in 1865 and was scaled with a 6 foot diameter, segmented into quadrants. His initial intention was to complete Lohrmann's work. However, upon reviewing his own work in April 1868, he was so dissatisfied with the level of detail that he decided to start afresh. Taking the current form, of 6 ft. in diameter, but split this time into 25 sections. This map is not a mere visualisation of the Moon's surface features, but an accurate representation of the positions and the sizes of all these features.

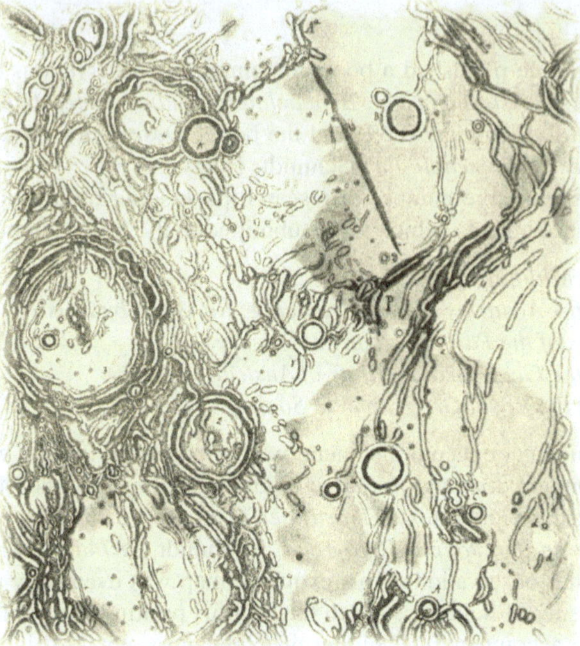

Fig. 16 Detail from Schmidt's Lunar Map. (From National Observatory of Athens)

Fig. 17 Schmidt's Lunar Map. (From National Observatory of Athens)

Fig. 18 Schmidt's Lunar Map. (From National Observatory of Athens)

Of special interest is the section on Linné, a crater in Mare Serenitatis that for years was cited as evidence for probable change on the Moon. Up to the year 1843 Linné was recorded as a small, deep crater, but in October 1866, Schmidt (1867b) re-observed it, noting that occasionally it also was merely a white spot. Pages 155 to 163 in his 1874 book *Charte der Gebirge des Mondes* … report individually his 200 observations of Linné from 1841 to 1874. His observations caused a scientific debate until it was declared that these changes were simply the result of an illusion caused by the illumination of the deep crater at different angles.

It is difficult today to understand just how difficult it was to make lunar observations like the ones Schmidt made. They demanded great attention and the synergy of different senses: hearing for measuring the time from ticking of a clock and the vision in order to accurately distinguish the moments when the edges of a formation on the surface of the Moon passed by the crosshairs of the micrometer, and all of this with the observer sitting in un*comfortable* positions behind the telescope's eyepiece.

Schmidt's map has been described by the great lunar cartographer Ewen Whitaker (1999:131), as "… the finest that was ever compiled without the aid of photographs."

As Richard Proctor (1886) explains in his book "*The Moon – Her Motions, Aspect, Scenery and Physical Condition*": "The labours of Schmidt, of Athens, must be regarded as altogether the most important contribution yet made to selenography."

Schmidt's fame had clearly spread wider than simply amongst the scientific community, for in the Jules Verne (1873) novel *Around the Moon*, we find the following reference: "We must be before-hand with Schmidt of Athens! He will leave nothing unnamed that his telescope can catch a glimpse of."

Fig. 19 Schmidt at his house in Athens. (National Observatory of Athens (copy from Prof. Ostwald Thomas collection))

On April 15, 1876 Schmidt's patron Simon Sinas passed away. His death raised the issue of the financial support of the Observatory, since the Greek State could not cover the expenses for Schmidt's salary and the operation of the Observatory. It was even proposed that the Austrian government should finance it, but fortunately Sinas' wife Iphigénie took over this burden, securing Schmidt's position and the continuation of his scientific activities.

Fig. 20 Simon and Ifigeneia Sinas who financed Schmidt and Athens Observatory. (From Bildarchiv Nationalbibliothek Wien)

Schmidt was awarded an honorary doctorate by the University of Bonn at their foundation festival, held in 1868. The RAS elected him an Associate on January 9, 1874, and he was the second recipient of the Valz Prize, awarded by French Academy of Sciences in 1878 for his selenographical work. A significant part of the Moon's nomenclature we use today is based on Schmidt's original nomenclature.

From 1871, Schmidt, who, due to age and distance, was unable to work regularly at the Observatory, requested the Berlin Academy of Sciences provide him with a small equatorial telescope. This was granted, and thus enabled him to continue his astronomical observations from the tower of his house, situated near Mount Lycabettus, right up until his death.

Fig. 21 Schmidt two years before his death. (From National Observatory of Athens)

Schmidt passed away in his sleep on the night of January 26 1884, after attending a social event at the German Embassy. He was only 59 years old. The King and Queen of Greece, all the Greek government, the local authorities, ambassadors, the university professors, students and thousands of Greek citizens attended his funeral. Many obituaries were written in the Greek press and in many scientific journals. His fame and his beloved low profile character persuaded the Greeks to consider him as one of their own.

Fig. 22 The Plössl equatorial refractor– Schmidt's main working instrument during his stay in Athens. (Courtesy Theofanis Matsopoulos)

The Schmidt Legacy

If we want to evaluate his work we may say, that he was a tireless observer of the heavens, who wanted to accumulate huge amounts of data and observations rather than just to improve his methodology and observational techniques. Many of his numerous published papers were actually short observational reports.

He never carried out modern Physical Astronomy (now called Astrophysics) partly due his isolation in a small country far from the main centres of scientific research at the time, partly due to his lack of formal education in physics, and partly because of his passion—one could almost say obsession—with the Moon. He did little to improve the instrumentation of Athens Observatory or to teach and inspire others in order to be his successors. Perhaps the main reason was the fact that he never learned Greek and thus his communication with the local scientific community was rather limited. Even his part-time assistant, Professor Dimitrios Kokkides who had studied in Germany failed to benefit enough from his experience. This can be deduced by Kokkides' little activity after taking over the Observatory as temporary Director, following Schmidt's death. We also may say that Modern Greek astronomy lost a great opportunity to flourish due to his presence for so many years in Athens. But in the history of astronomy he holds a significant position. There are craters named after him on the Moon and Mars and there is also a crater named after Sinas on the Moon to remind us of the tale we have briefly presented here.

	Coordinates	Diameter
Crater Schmidt on the Moon	1.0°N, 18.8°E	11 km
Crater Schmidt on Mars	72.3°S, 78.1°W	201km
Crater Sinas on the Moon	8.8°N, 31.6°E	12 km

Alternatives to *Der Mond*

Although Schmidt's map was well advanced in 1868, it was not completed until 1874. However, Athens Observatory could not afford to publish it. Even Fellows of the Royal Astronomical Society made enquiries in an attempt to assist with costs. However, terms could not be agreed. Whilst displaying his map at Berlin Observatory in 1874, such was the admiration amongst the scientific community, that the Crown Prince himself urged the Prussian Government to fund the publication. Four years later the General Staff Office, under the leadership of Count von Moltke produced the photolithographed map. Twenty-five sections, compiled using over one thousand drawings and three thousand height measurements, the map, made up of 25 sections, was finally published in 1878 together with a descriptive companion book (*Charte des Mondes nach eigenen Beobachtungen in den Jahren 1940-1874*). Thirty-four years after Schmidt had begun his quest, he had finally produced the most detailed hand drawn map of the Moon that the world has ever seen. Still in use well into the following century, it was never bettered.

Following the publication of Beer and Mädler's lunar map (*Mappa Selenographica*, 1836) and the accompanying book (*Der Mond nach seinen komischen und individuellen Verhältnissen oder Allgemeine vergliechende Selenographie*, 1837) it is fair to say that very little lunar interest existed – excepting that of Schmidt. Presumably most observers believed that Beer and Mädler's map and book, described most of what could ever be known about our satellite. However, reviewing the articles that appeared in both *Monthly Notices of the Royal Astronomical Society* and *Astronomische Nachrichten*, it can clearly be seen that Schmidt's announcement of the obscuration and changes of appearance of the crater Linné, reignited astronomers' interest.

In 1852 the British Association formed a Moon Committee consisting of Lord Rosse, Rev. Dr. Romney Robinson and John Phillips to compare the lunar surface with that of the Earth. Later John Phillips would also enrol Warren De la Rue (early lunar photographer) and Thomas William Webb into the observational program.

James Nasmyth (inventor of the Steam Hammer) and James Carpenter (Greenwich Royal Observatory) published a book about the Moon (*The Moon Considered as a Planet, a World and a Satellite*, 1874). From its preface however, we can see that this book is not an all-encompassing treatise about the Moon:

> Much valuable labour has been bestowed upon the topography of the Moon, and this subject we do not pretend to advance. Enough has also been written for the benefit of those who desire an acquaintance with the intricate movements of the Moon in space; and accordingly, we pass this subject without notice. But very little has been written respecting the Moon's physiography, or the causative phenomena of the features, broad and detailed, that the surface of our satellite presents for study.

After explaining the volcanic formations of the Moon, Nasmyth and Carpenter then do indeed proceed to describe the general topography of the Moon, including a very rough map indicating approximately 240 visible features. Notable with the later editions of this book, are the photographs of excellent plaster models (given as examples of various lunar surface topography).

In 1876, Edmund Neison finally published a book that the English-speaking world had been waiting for. Based upon *Der Mond* of Beer and Mädler, although not a direct translation, it certainly followed the same format. *The Moon and the Condition and Configuration of its Surface* was an extensive tome, not only covering all physical characteristics and lunar history, but also descriptions of features referencing both Mädler and Schmidt's published data – this is almost certainly why the English-speaking community then found no need to translate these German works – Neison had conveniently incorporated them into his work. What Neison included (and was noticeably absent from the *Der Mond* books, were reference maps. Neison collated his own and other lunar observations (notably by Rev. T.W. Webb) over a period of 8 years, and interspersed the book with 22 pages of maps (using a scale of 24 in. to the Moons diameter). This clearly made the book more useful,as the hobbyist astronomer in the late 1800's probably would not have been concerned about the smaller scale, nor the lack of detail, when compared with Schmidt's work. Neison and Schmidt exchanged some feature names through their correspondence, and there were few name clashes, despite Schmidt's map still being another two years in the making. Neison even had his book and map printed separately in German.

The British Association for the Advancement of Science formed a committee in 1874 to stimulate lunar research. Led by William Radcliffe Birt (who translated much of Schmidt's work from *Astronomische Nachrichten*, for inclusion in the *Monthly Notices of the Royal Astronomical Society* and *The Observatory*), they tasked themselves with the construction of the largest scale map of the Moon yet – 200 in. (compared to 74 in. of Schmidt and 34 in. of Beer and Mädler). Unfortunately, Birt who was also secretary of the Selenographical Society, died in 1881. Together with his friend John Lee, they added 85 new names to features – features which Schmidt had already named with alternatives, despite Birt and Schmidt being in correspondence over such issues. Schmidt stated that he was unaware until his work was nearly complete, of the many new names proposed by the English observers (sometimes referred to as the 'Lunar Committee' or the 'Moon Society of London'). The Editor, Edmund Neison (whose actual name was Edmund Neville Nevill), at about the same time resigned, to take up the role of Director of the Natal Observatory in South Africa. The map was never completed.

Thomas Gwyn Elger continued the work with The Lunar Section of British Astronomical Association, and published a book (*The Moon*, 1895) with a standalone 18 in. in diameter map 2 years before his death in 1897 and before the use of photography changed selenography forever.

With the employment of photography, visual maps of the Moon were completely surpassed, since photographs provide objective data that can be processed later with better accuracy. During the 1890s and 1900s images of the Moon were taken of sufficiently high quality and detail to form the basis of the earliest photographic atlases of the Moon.

In 1897, Edward Singleton Holden began issuing in serial form, the Lick Observatory *Atlas of the Moon* compiled by Ladislaus Weinek from photographs obtained at the Lick and Paris Observatories. Only 19 sheets of reproduced photographs, out of the 60 originally intended were ever published; they were of low resolution and poor quality.

In 1903 the Harvard astronomer, William Henry Pickering (1858–1939) became the first to publish a complete photographic atlas of the Moon with the appearance of his *The Moon. A Summary of the Existing Knowledge of our Satellite with a Complete Photographic Atlas* (Annals, Harvard College Observatory, Volume 51, 1903). His photographs were taken in an eight month period from 31st December 1900 to 31st August 1901 from Mandeville, Jamaica using a horizontal refractor of 12-in. aperture and 135 feet focal length. Pickering divided his atlas into sixteen areas, with photographs taken of each area but under five differing illuminations by the sunlight.

Although the photographs he obtained were not of the highest quality, the fact that the same features were imaged during differing lunar phases was a very useful feature of the atlas. It is well known that a lunar feature under differing illumination provide very differing aspects, which are often difficult to identify without photographic help.

However the first true photographic atlas of the Moon was the work of Moritz Loewy (1833–1907) and Pierre Henri Puiseux (1855–1928) of the Paris Observatory. In 1894 they took the first photograph of the Moon using the Observatory's 23.6-in. Equatorial Coude Refractor that would be included in their monumental *L'Atlas Photographique De La Lune* (Observatoire de Paris, 1903).

Their work was published in twelve parts. Each part relates to a specific area of the Moon, and contains high quality photographs of the region, as well as a description of the major lunar features, craters, mare, mountain ranges etc. present. A general index of features was also published as the thirteenth part of the atlas.

The photographs of Loewy and Puiseux were to remain unsurpassed in quality for the next decades until G.P. Kuiper published the *Photographic Lunar Atlas* (The University of Chicago Press, 1960) and those taken by the Lunar Orbiter Probes in the 1960s, although Francis Gladheim Pease came very close with his images of 1919, taken with the 100-in. Hooker Telescope at Mount Wilson. In 1964, the Lunar Orbiter program was initiated, as a series of five unmanned Lunar Orbiter missions launched by the United States from 1966 through 1967. All five of the Lunar Orbiter missions were successful, and 99 percent of the Moon was mapped from photographs taken with a resolution of 60 metres or better.

In 1971 the images obtained from the Lunar Orbiters were used to create *LOPAM – the Lunar Orbiter Photographic Atlas of the Moon* compiled by David Bowker and J. Kenrick Hughes. This work is now considered to be the definitive reference material on the global representation of the Moon.

Fig. 23 The original building of Athens observatory (now called Sinas building) as it is today. It is now a museum. (Courtesy Theofanis Matsopoulos)

Acknowledgements This introduction was written with extensive contributions from Nicolas Matsopoulos at the National Observatory of Athens.

References

Argelander, F.W.A. (1859) *Bonner Beobachtungen Bd. III-V.* Berlin, Berlin University Observatory.

Ashbrook, Joseph (1984) *The Astronomical Scrapbook*, edited by Leif J. Robinson. Cambridge(Massachusetts), Sky Publishing Corp.

Barabas, Miklos (1866). *HonpolgarokKonyve, Volume 1*. Budapest.

Batten, Alan H. (1991) Argelander and the Bonner Durchmusterung. *Journal of the Royal Astronomical Society of Canada*, 85, 43-50.

Baum R. (2014) "Beer, Wilhelm." In: Hockey T. et al., pp. 107-108.

Beer W., &Mädler J. (1834) *Mappa selenographica : totam lunae hemisphaeram visibilem complectens observationibus propriis : quatuor sectionibus constructa et delineate.* Berolini, S. Schropp.

Dean's Report for the Year 1880.University of Athens, Athens (1881).

Dean's Report for the Year 1881.University of Athens, Athens (1882).

Dobbins, T.A., & Sheehan, W.P.(2001).*Epic Moon. A History of Exploration in the Age of the Telescope.* Richmond, Willmann-Bell.

Elger, T.G. (1895). *The Moon: A Full Description and Map of its Principal Physical Features.* G. Philip & son, London.

Farrington, O.C. (1925) *The Moon*. Chicago, Field Museum of Natural History (Geology, Leaflet 6).

Hahn, J.G. von (1865) *Die Ausgrabungen auf der homerischen Pergamos : in zwei Sendschreiben an Georg Finlay*. Englemann, Leipzig.
Hahn, V.G. von (1867) *Reise durch die Gebiete des Drin und Wardar im Auftrage der Kaiserl. Akademie der Wissenschaften, unternommen im Jahre 1863.* Band I-II, Wien.
Hockey, T. et al. (2014) *Biographical Encyclopaedia of Astronomers*. Second Edition. Springer International Publishers, Cham (Switzerland).
Holland, J. (2014) "Rümker, Christian Karl Ludwig." In: Hockey T. et al., pp. 900–901.
Joeveer, M. (2014) "von Mädler, Johann Heinrich." In: Hockey T. et al., pp. 723-724.
Kokott, W. (2007) "Benzenberg, Johann Friedrich." In: Hockey T. et al. (eds) *The Biographical Encyclopedia of Astronomers*. Springer, New York, NY, p.111.
Kuiper, G.P. (1960) *Photographic Lunar Atlas*. The University of Chicago Press, Chicago.
Laios, G. (1962) *The Athens Observatory*. Unpublished Documents. Athens (In Greek).
Lardner, D. (1867) *Handbook of Astronomy*. James Walton, London.
Loewy, M.M., & Puiseux, P.H. (1903) *L'Atlas Photographique De La Lune*. Observatoire de Paris, Paris.
Markkanen, T. (2007) "Argelander, Friedrich Wilhelm August." In: Hockey T. et al. (eds) *The Biographical Encyclopaedia of Astronomers*. Springer, New York, NY, p.58.
Nasymth, J., & Carpenter, J. (1874) *The Moon Considered as a Planet, A World and a Satellite*. John Murray, London.
Nöggerath, J. (1857) *Geschwindigkeit der Erdbebenschwingungen*, In: "*Das Erdbeben vom 29 Juli 1846 im Rheingebiet und den benachbarten Ländern*", Henry and Cohen, Bonn.
Matsopoulos, N. (2000) *Astronomy in Modern Greece*, 1700-2000. National Observatory of Athens, Athens (In Greek).
Neison, E. (1876) *The Moon and the Condition and Configuration of its Surface*. Longmans, Green & Co., London.
Pickering, W.H. (1903) *The Moon. A Summary of the Existing Knowledge of our Satellite with a Complete Photographic Atlas* (Annals, Harvard College Observatory, Volume 51).
Proctor, R. (1886) *The Moon – Her Motions, Aspect, Scenery and Physical Condition*. Longmans, Green & Co., London.
Schröter, Johann Hieronymus (1791) *Selenotopographische Fragmente zurgenaueren Kenntnis der Mondfläche, ihrererlittenen Veränderungen und Atmosphäre sammt den dazu gehörigen Specialcharten und Zeichnungen.*Lilienthal & Göttingen: Auf Kosten des Verfassers.
Theodossiou, E.Th.,Manimanis, V.N., & Mantarakis, P. (2007) Demetrios Eginitis: restorer of the Athens Observatory. *Journal of Astronomical History and Heritage*, 10, 123–132.
Verne, Jules, 1873. *Around the Moon*. London, Ward, Lock and Tyler.
Whitaker, E.A. (1999). *Mapping and Naming the Moon*. Cambridge, Cambridge University Press.
Weinek, L. (1901) *Photographischer Mond-Atlas : vornehmlich auf Grund von focalen Negativen der Lick-Sternwarte im Massstabe eines Monddurchmessers von 10 Fuss*

Bibliography

Clerke, M.A. (1887) *A Popular History of Astronomy during the Nineteenth Century. 2nd Edition*. Edinburgh, Adam & Charles Black.
Cunningham, C.J. (2014) Schröter, Johann Hieronymus. In: Hockey et al., pp. 1030-1031.
Dunkin, E. (1885) Johann Friedrich Julius Schmidt. *Monthly Notices of the Royal Astronomical Society*, 45, 211–212.
Haidinger, Wilhelm von, (1858) *Mittheilungen von Herrn J. F. Julius über Feuermeteore*. Karl Gerold, Wien.
Pflicht, C.A. (2014) Lohrmann, Wilhelm. In:Hockey et al., pp.704-705.

Kreutz, Heinrich Karl Friedrich, (1884) Johann Friedrich Julius Schmidt.*Astronomische Nachrichten*, 108(2577), 129–130.
Lynn, W.T. (1884) Dr. Julius Schmidt (obituary).*The Observatory*, 7, 118–119.
Lynn, W.T. (1875) Johann Heinrich von Mädler (obituary).*Monthly Notices of the Royal Astronomical Society*, 35(4), 171-175.
Tampakis, K. (2014) Onwards facing backwards: the rhetoric of science in nineteenth-century Greece. *The British Journal for the History of Science*, 47, 217–237.
The Hellenic Archives of Scientific Instruments: National Observatory of Athens (http://www.eie.gr:80/hasi/hasi/institut/noa.html) [HASI].

Schmidt Bibliography

[Note, below, that where Schmidt published more than one paper in the same year, the papers have been listed a, b, c, etc., alphabetically by title, so that each paper may be easily identified and if necessary mentioned in the text.]

Schmidt, Johann Friedrich Julius (1852a) *Beobachtung der totalen Sonnenfinsterniss von 28sten; Juli, 1851, zu Rastenburg in Ostpreussen.* Bonn.
Schmidt, Johann Friedrich Julius (1852b) *Resultate aus zehnjährigen Beobachtungen über Sternschnuppen*. Georg Reimer, Berlin.
Schmidt, Johann Friedrich Julius (1854a) "*Berechnung des Durchmessers von Mondholen.*" In: *Annalen Der Physik*, 168(6), 324-333
Schmidt, Johann Friedrich Julius (1854b) *Das Relief der sichtbaren Halbkugel des Mondes, angfertigt von Th. Dickert.* Carl Georgi, Bonn.
Schmidt, Johann Friedrich Julius (1856a) *Das Zodiacallicht. Uebersicht der seitherigen Forschungen nebst neuen Beobachtungen ueber diese Erscheinung in den Jahren 1843 bis 1855.*Schwetachke, Braunschweig.
Schmidt, Johann Friedrich Julius (1856b) *Der Mond: Ein Überblick über den gegenwärtigen Umfang und Standpunkt unserer Kenntnisse von der Oberflächengestaltung und Physik dieses Weltkörpers.* Johann Ambrosius Barth, Leipzig.
Schmidt, Johann Friedrich Julius (1856c) *Die Eruption des Vesuv im Mai 1855, nebst Beiträgen zur Topographie des Vesuv, der phlegräischen crater, Roccamonfina's und der alten Vulkane im Kirchenstaate, etc.* Friedrich Manz, Wien [printed] und Olmütz.
Schmidt, Johann Friedrich Julius (1856d) *Neue Höhen-Bestimmungen am Vesuv, den phlegräischen Feldern, Roccamonfina und im Albaner-Gebirge :nebst Untersuchungen über die Eigenschaften und Leistungen des Aneroid-Barometers.* Hölzel, Wien [printed], Olmütz.
Schmidt, Johann Friedrich Julius (1857a) *Resultate aus elfjaehrigen Beobachtungen der Sonnenflecken.*Hölzel,Wien [printed] und Olmütz.
Schmidt, Johann Friedrich Julius (1857b) "*Geschwindigkeit der Erdbebenschwingungen*" In: Nöggerath's *Das Erdbeben vom 29 Juli 1846 im Rheingebiet und den benachbarten Ländern.* 28-37.
Schmidt, Johann Friedrich Julius (1858a) *Charte der Verbreitung und Intensität des Erdbebensam am 15. Jänner 1858.* Wien.
Schmidt, Johann Friedrich Julius (1858b) "Untersuchunguber das Erdbebenvom 15 Juni 1858." In: *Mittheilungen der Wiener Geographischen Gesellschaft.* Wien.
Schmidt, Johann Friedrich Julius (1858c) *Untersuchungen ueber die Leistungen der Bourdon'schen Metallbarometermit Hinweisung auf den Nutzen dieser Instrumente fuer die Marine.* Hölzel, Wien [printed] und Olmütz.
Schmidt, Johann Friedrich Julius (1861) *Beiträgezur Physikalischen Geographie von Griechenland.*Athens, Observatoire d'Athènes (Publications de l'Observatoired'Athènes, 11me série, tome 1).

Schmidt, Johann Friedrich Julius (1863) *Astronomische Beobachtungen über Cometen.* Karl Wilberg, Athens.

Schmidt, Johann Friedrich Julius (1866) *Ueber Rillen auf dem Monde.* Johann Ambrosius Barth, Leipzig.

Schmidt, Johann Friedrich Julius (1867a) *On the 23rd of January 1867 Earthquake of the Kephalonia.*Athens (in Greek).

Schmidt, Johann Friedrich Julius (1867b). Ueber den Mondcrater "Linne". *Astronomische Nachrichten*, 68(1631), 365–366.

Schmidt, Johann Friedrich Julius (1868). *"Mittlere Oerter von 110 Nebeln für 1865".Astronomische Nachrichten*, 70(1678), 343–352.

Schmidt, Johann Friedrich Julius (1869) *Astronomische Beobachtungen über Meteorbahnen und deren Ausgangspunkte.* Athens, Observatoire d'Athènes (Publications de l'Observatoired'Athènes,1re série,tom. 2,1869).

Schmidt, Johann Friedrich Julius (1874a) *Charte der Gebirge des Mondes – nach eigenen beobachtungen in den Jahren 1840-1874.*Reimer, Berlin.

Schmidt, Johann Friedrich Julius (1874b) *Vulkanstudien– Santorin 1866 bis 1872. Vesuv, Bajae, Stromboli, Aetna.* Carl Scholtze, Leipzig.

Schmidt, Johann Friedrich Julius (1875)*StudienüberErdbeben.*Carl Schöltze, Leipzig.

Schmidt, Johann Friedrich Julius, Lohrmann (1878) *Mondcharte in 25 Sectionen und 2 Erläuterungs tafeln. Mit Erläuterungen und Selenographischen Ortsbestimmungen, unter Benutzung des von den Herren F. W. Opelt und M. Opeltrevidirten und ergänzten Materials, zu gleichals Supplement zu Lohrmann's Topographie der sichtbaren Mondoberfläche Abth. 1, herausgegeben von Dr J. F. J. Schmidt.*Barth, Leipzig.

Schmidt, Johann Friedrich Julius (1881) *Studienüber Vulkane und Erdbeben.*Alwin Georgi, Leipzig.

Schmidt, Johann Friedrich Julius (1884) *Description Physique d'Attique.Meteorologieet Phenomenologie*. Librairie de Charles Beck, Athens.

Schmidt, Johann Friedrich Julius (1911) "*Zur Meteorologie von Athen: Witterungs auf zeichnungen 1863-1879; Messungend Radien d. Mondhalo von 22° Nordlicht beobachtungen.*" Edited by Karl Knoch, Springer, Berlin.

Translator's Note

As part of translating a piece of work, one must inevitably make some decisions regarding the content and the audience of the translation. The following lays out a few directions taken with this book.

Measurements

Schmidt measured all that he observed—heights, depths and distances. Today, we would probably expect these units to be recorded using the metric system of kilometres. In the age of Schmidt's work, one would probably expect these to be recorded in the older imperial style of miles and feet. However, Schmidt recorded everything using a measurement system of the Toise. I wouldn't expect many people to even have heard of such a system, but during the 1850's within the European Scientific community, it was certainly the unit of choice. However, there were several versions of the Toisen. For instance, the Vienna Toise was not the same length as the Paris Toise or the Cologne Toise (see https://www.sizes.com/units/toise_local.htm). I have left all data in the Toisen measurement system. The conversion is provided by Schmidt himself in the Annotations chapter at the end of the book—see Annotation 3. Approximately, 1 Toise = 2 metres.

Name Changes

Throughout history, selenographers have both created new names and amended the names of existing Lunar craters (which are named after individuals), sometimes to ingratiate themselves to their sponsors, or sometimes simply because they perhaps felt somebody needed the recognition. Schmidt was not immune to this behaviour. The following is a list of crater names that are encountered in Schmidt's original text, followed by the current official name.

Aristardi	=	Aristarchus
Buchius	=	Byrgius
Cap. Huygens	=	Mons Huygens
Bulliald	=	Bullialdus
Landsberg	=	Lansberg
Burkhardt	=	Burckhardt

The IAU is responsible for the official naming that we currently use.

As you may see from the list, some of the names are slight changes in spelling. However, some features have had the name changed completely. If you search the Moon for Mare Kästner (home of crater Kästner), using a modern atlas, you'll fail to find it.

Schröter introduced this named feature, Mädler subsequently dropped it (or failed to use it and simply chose to use it for the named crater instead) and Schmidt chose to continue using it. However, the name changed completely to Mare Smythii (The Sea of Smyth) circa 1863 (Report of the 1863 Meeting of the British Association, pages 7–9 of the Notes Section).

Within the text I have appended the official IAU names within square brackets following the original name—although within the tables I have used just the original name so as not to distract the reader from the data represented.

Perhaps unsurprisingly, locations and features here on Earth have also undergone name changes since Schmidt wrote his book. I have changed names to their modern incarnation where appropriate. One such example is the Sandwich Islands—now known as Hawaii.

One town also mentioned is Dorpat. This was the home (and observatory) used by Mädler. Today this Estonian town is more commonly known as Tartu, although some people still refer to it as Dorpat. See https://www.tahetorn.ut.ee/en/content/history-old-observatory. More subtle differences occur within the text. Mount Kintchinjinga (which may be a misspelling or German derivative) is better known today as Mount Kangchenjunga.

The titles of books and articles have retained their original language title rather than translating them.

Schmidt Revelations

The book content leads you to glean a little more insight into the life of Schmidt. He was almost certainly well-travelled for the age. Details provided of height measurements would have entailed extensive trips to the Canary Islands, equatorial South America, The Azores and especially Italy. The numerous details of Italian volcanos would certainly have needed several expeditions. Schmidt was very well read—he would have needed access to many books and articles to provide the references included.

Opinions about living beings on the Moon and on the planets (Chapter 27) demonstrate his desire to think more about science and the possibilities that it could reveal. Although an impeccable scientific mind, Schmidt showed he was not just made of cold scientific logic. Chapter 28, *A day and a night on the Moon*, reveals a more expressive, imaginative and almost poetic way of thinking about science.

During the book's research it became apparent how famous Schmidt was, especially within German speaking Europe. He was well connected and partook in correspondence with other well-known scientists of the day, notably Alexander von Humboldt. Many newspapers of the day carried advertisements for his books and lectures.

The number of Obituaries and newspaper reports surrounding Schmidt's funeral— even English-language newspapers—reaffirm his fame and underline how surprising it may be that he has since not only been forgotten by the general public but also by a large portion of the scientific community.

Other Revelations

In Chapter 15, some readers may be confused by the use of Mount Kangchenjunga as a comparison. This is because at the time, it was considered to be the highest mountain on Earth. Mount Everest was only officially announced as the highest in 1856 as a result of the British Great Trigonometric Survey of India.

On a personal note, whilst translating this book, I did encounter geological terminology previously unknown to myself. Now I know Strike and Dip, and that Neptunic rock (sedimentary rock) is rock formed under the sea, and that Plutonic rock (metamorphic rock) is rock formed at considerable depth by crystallization of magma or by chemical alteration.

Further study is encouraged within the Annotations chapter. One such snippet of information is that of [74] Captain Denham on the H.M.S. Herald, who measured the deepest part of the ocean at 7706 fathoms (over 8 miles). Although almost certainly a flawed measurement, this would have been a widely reported scientific breakthrough of the time.

The Annotations chapter certainly contains as much additional information as the contents of the book, and will provide a wealth of material for further research by the reader.

Translated by Stephen Harvey

THE MOON

An overview
of the present extent and the standpoint of our knowledge
of the surface structure and physics
of this heavenly body.

From J. F. Jul. Schmidt,
astronomer of the Observatory of the Prelacy of Ritter von Unkrechtsberg
to Olomouc.

In addition to two coloured lithographic plates and several woodcuttings
printed in the text.

Leipzig.
Published by Joh. Ambr. Barth.
1856

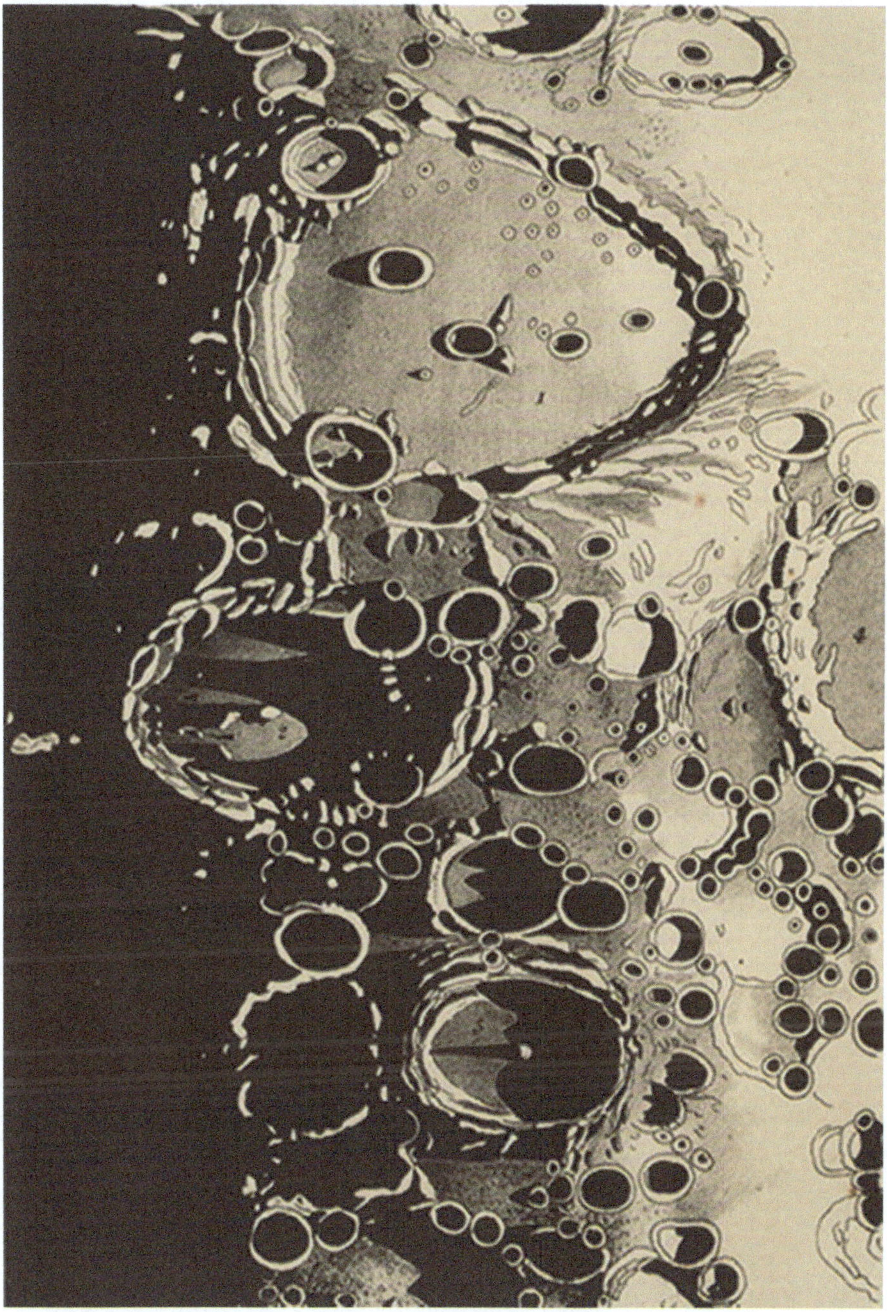

Shadows across the crater landscape of the Moon around Tycho in the setting Sun

Preface

This Work on the mountains of the Moon, which I present here not only to the astronomers, but preferably to the geologists, as well as to all admirers of the natural sciences, is to be regarded as the forerunner of a larger work of selenography, in which the works of LOHRMANN will publish the Moon-mountains at the same time as my own observations, begun in 1840, on the surface of our Earth satellite.

In view of the large volume of material and the difficulty of treating a strictly scholarly work in a popular manner, it seemed very appropriate to me, to compile all previously known results of telescopic observations of the lunar surface, to explain this easily, and at least so far to mention the essentials of the movement, mass, size and illumination of the Moon, so that one with little understanding of Astronomy may still understand.

But it still has the special purpose of pointing out that a careful study of the shape of the Moon-mountains may become of importance to geology, inasmuch as it is at some time a question of certain similarities between the mountain forms of the Earth and its satellite, and a comparative in which we examine the effects of immense forces which have given their configuration to the surfaces of two adjacent heavenly bodies.

There remains no hope of ever knowing the natural constitution of the constituent parts of which the mountains of the Moon are composed, and, as in geology, we also find ourselves completely at a loss to distinguish metamorphic formations from igneous rock mountains of our companion, so we are confronted with the substantive investigation and comparison of the forms and the dimensions, which alone will lead to a useful result.

Our knowledge of the surface of the bodies in question is regardless of the wealth and importance of individual facts, notwithstanding the completeness of the survey, which we have in particular of the mutual position, size, and height of the mountains of one half of the Moon facing us, far too fragmentary, however, that a rigorous comparison and reasoned conclusion could now seem permissible.

But it is also just the beginning to be made and the research to open a new path.

Geology is still too young a science to provide sufficient information about all forms of terrestrial mountains.

Its' beautiful, often grand, and splendid results have not been able to determine the limits within which the water or fire formed the presently solidified forms; it has not yet led to the establishment of a completely general, comprehensive definition of volcanism.

Regarding volcanism, A. v. HUMBOLDT 's understanding of the reaction of the interior of a planet to its surface is generally understood, so, under this effect, with reference to the Earth, all the metamorphic forms are also included in so far as they have been brought from their originally essentially horizontal position into a changed position, or completely destroyed, or destroyed only by the formerly molten rock types driven up from within the Earth and the volcanoes appear only as scattered phenomena in the places where the breaking forces formed crater-shaped gullies, elevation craters with or without later break-out cones, or high mountains closed at the top, whose rock more or less clearly betrays the formerly fiery-liquid state.

Thus volcanism has altered, to a lesser extent, only the mathematically rigorous shape of the Earth, the surface of a rotational spheroid, whether it devastated the already rigid old surface, or bringing new materials of a completely different composition from the depths of the Earth to light, which were later broken here and there by crater-mountains, by numerous, now partly active, partly extinct volcanoes.

According to this view, which takes no account of the external structure and position, as well as of relative age, but of the nature of the masses of rock, it may be said that the surface of the Moon as we now find it had been formed by volcanism, without assuming that the Moon must have volcanoes similar to Earth, and that it is in fact directly comparable with those of the Earth in terms of its mountains.

But we may ask whether between the volcanic craters and the gigantic ring-mountains of the Moon, between the volcanic craters of the Earth and its companion, we may be able to show the existence of such a consistency of the structure, which is more or less the same similar type of origin; we may at last examine whether or not we have a dependence of the mountain directions with the rotational axes for both heavenly bodies.

These are the large and general features, and if the expression is permitted, the cosmic characteristics from which we try to guess the ancient history of these bodies.

If there is a great danger of being led astray by seemingly striking similarities of form, then the existing scientific material is sufficient to be able to vigorously oppose a superficial approach based on hypotheses; and if I especially point out that I only intend to venture a new step in this region which has hardly been entered, and try to communicate more the necessary preparatory work than my own views on the results to be drawn from these, I may perhaps hope that with this attempt, whose actual execution in particular is reserved for the larger works, will be accepted with some leniency.

It is very difficult for the individual to have all the necessary knowledge of the geological writings, of the land and sea-charts, however, these circumstances could not deter me of my plans.

In my opinion that an independent study in the great outdoors must be combined with the study of books, if one does not want to become completely dependent on random opinions in geology, I have during my seven-year stay in Bonn, made more frequent trips besides other purposes to see many mountains for themselves and to investigate them more closely of course, I turned my attention mainly to the volcanoes, especially the craters of the Eifel, and more recently to the famous fire mountains of Italy. Many accounts, which seem to me to be desirable for my endeavours, are either not found in the books, or are very incomplete.

In numerous geological works of varying value, one misses a sufficient account of the general formations, the dimensions, and the slopes of our volcanoes, and many values, when combined from different sources, often reveal that they are based only on very crude estimates; and that it is not uncommon to distinguish either the toisen from the meter, nor the English from the French foot.

Some of the mountains have been described with great care in the mineralogical sense, but rarely is such a thorough description of their heights and formations, as we have in the excellent treatises on the Rucu-Pichincha, the Pics of Tenerife, Palma, Vesuvius, and Etna. But these forms alone are the ones which the observer of the Moon has to compare. We can judge the mountains of our satellites only by their shape and by their dimensions.

Continued measurements and observations on the inclination angles of the mountainous areas, whether on single elevations, or on the ramparts of large ring mountains, constructions of the mountain rocks, determined from the outline of their shadows, though already in this work, to a much greater extent, are presented only in the greater work, in order to facilitate the comparison of the mountains of both bodies, and, in conjunction with the topographic representations, of the future, a true, much more comprehensive picture of the present condition the surface of the Moon.

The much-mentioned large-scale Selenography will not appear until after two or three years.

In the meantime, in order to stimulate a more general interest in the Moon, after almost 18 years, when MÄDLER's highly-valued works have been published, almost no one has turned his attention to our neighbour's world, and finally the very large number of fallacies and groundless hypotheses about the Moon I have, in my studies in Olomouc, compiled this smaller, generally comprehensible book, which, almost independently of all the views and hypotheses which have since been given on this subject, is supported entirely by my own experiences and observations, but where it seemed necessary, to the work of previous Selenographers.

I have from frequent contact with persons of all classes and of very different degrees of education, the often-confirmed view that, with very few examples excepted, there is nowhere near a correct understanding of the present state of our knowledge of the Moon.

All the questions which I had to answer for ten years were almost exclusively related to the inhabitants, and, when it came up, regarding the mountains of the Moon, as well as to the methods of measuring them.

MÄDLER's exquisite illustrations seemed to be forgotten in larger circles, and names like TOBIAS MAYER, SCHRÖTER, and LOHRMANN are almost unknown.

But the recollection of a book published some twenty years ago, in which miraculous observations of the Moon people were made by the famous Sir JOHN HERSCHEL, was still present, albeit often accompanied by just doubts about the credibility of a writing which, to a lesser degree, but in other respects very absurd, could only have been calculated for a great deception of the public.

The celestial body, however, which has already inspired the old human race to curious, sometimes to more serious questions, to questions which for thousands of years kept the same vitality of interest, to questions that have retained the same liveliness of interest for millennia without which one had only sought to investigate only the laws of the motion, quite apart from their astronomical importance, it deserves to be chosen preferentially as the object of repeated observations and special studies.

Unfortunately, we have only too many hypotheses, and the confusion of the views on nature on the Moon is as great as the discredit, in which of these celestial bodies, as far as the observation of its surface is concerned, many scholars have come, admittedly in a very unfounded manner.

MÄDLER was the only one in this century after LOHRMANN, who sought to eliminate this mis-information in the most dignified manner, by making new drawings and measurements, and avoidance of uncertain and arbitrary hypotheses.

In the same direction, in a somewhat broader sense, I have been concerned with the Moon since the beginning of my astronomical activity, and these observations, notwithstanding the lesser favour given to them, for fifteen years, without having laid more than the first foundations for a venture which, often hindered and always affected by the sad climate of Central Europe, will give me the old undiminished interest for a long time.

May I also presuppose that some astronomers will not approve of these endeavours, because they do not correspond to the direction of today's astronomy, so I am completely relieved of the fact that the lack of recognition on this front is sufficiently compensated by the participation of the great number of those , who find it quite in order when, in the course of every quarter of the century, at least one of the many observing and calculating astronomers is concerned somewhat with the mountains of the Moon.

It is not for me to express an opinion on the relative value of different directions in the exact sciences, but one should not misinterpret it if I take the view and act accordingly, that those aspirations for the knowledge of nature are also considered noble and the eternal activity of the human mind, whose results cannot always be expressed in numbers or judged by their probable error.

Rome, 23rd March 1855

Contents

Print Errors

Which you'll want to modify in advance.

Page 10 Line 6 from the top left her instead of him.
Page 13 Line 16 from the top left vermaass instead of vermass.
Page 73 Line 10 from the bottom gewährt instead of gewährte.
Page 87 Line 14 from top left Grunde statt Gründe
Page 102 missing in Fig. 3 in the crater 1 left and right of h the letters p and q.

Introduction

General recollections about the Orbit and about the size of the Moon

Although it is only intended in this document to describe the surface of the Moon, the shape and height of its mountains, it seems advisable to touch upon those astronomical conditions which are necessary for the understanding of selenography, which is itself outlined in brief.

I refer everywhere to the values quoted by MÄDLER in his great work [1] , and in the notes give recent results or individual selenographic fragments, which may perhaps be of use to one who is particularly concerned with the study of the Moon.

At the observatories one observes the Moon, in order to know the shape and the changes of its course with the help of the calculations from the exact recording of its positions in the sky.

The very difficult, extremely complicated problem of the true motion of the Moon, even in our days, has not yet found its perfect solution, which meets the exact requirements.

We owe much to the investigations in this direction, with a very profound knowledge of the orbit of the Moon and of those irregularities of its motion which, affected by the ever-changing action of the Sun and the planets, have been designated by astronomers with the term "perturbations."

The measurement in association with the calculation shows the apparent and the true size of the Moon; from the influence exerted by the gravity of the Moon on a certain movement of the Earth's axis, or on the ebb and flow of the sea, one concludes the mass of our satellite, and if these values are determined, then without difficulty we are able to conclude the mean density of matter, what the Moon is made of, as well as the gravity on its surface.

Only the most intimate union of observational art with applied higher mathematics is able to defeat the numerous obstacles which oppose the complete solution of the problem of the motion of the Moon.

This study, which belongs to the astronomer, completely excludes the study of the surface condition of the Moon.

The purely astronomical observation of our satellite is exclusively concerned with specifying precisely its apparent location at any time, and since the center of the Moon, understandably, cannot always be seen, the position of the Moon's edge, which is always visibly illuminated, must be determined in the sense of the right ascension and declination, or in other words, by observation, one should find the apparent location of the Moon in relation to the general graticule of the sky.

The understanding of the same must be assumed to be known here.

But since, because of the mostly incomplete illumination of the lunar disc, one can only observe the eastern or western, the northern, or the southern edge, it follows that a precise knowledge of the apparent diameter of the Moon is necessary.

The opportunity to determine that diameter directly with the help of a micrometre is rare; It is determined by observation of the passage of the Moon before the Sun or by fixed stars, but the whole complication of the motion of the Moon is presumed to be known.

If the apparent diameter is sufficiently well known, any observation of two edges may be reduced to the centre of the Moon, and the place of the centre of the Moon shall be obtained, as it appears to the observer on the surface of the Earth.

It can therefore be said that the astronomer's entire work, as far as the Moon is concerned, is confined to:

"at the time when this star is in the meridian of its observatory, to determine the right ascension of the eastern or western, and the declination of the northern or southern edge of the Moon."

Everything else that the theory requires is a matter of calculation.

For our purposes, it suffices to explain for a part of our readers the chief results of the orbital time, parallax, size, density, and libration of the Moon, and to casually recall for another part of them.

Chapter I
Orbital Period of the Moon

The Moon moves in an ellipse around the Earth itself moving in space, in the same common direction from west to east, which is characteristic of all known planetary bodies of the solar system, with the exception of the Uranus satellites, and likewise many comets.

The plane of its orbit inclines towards the ecliptic (orbital plane of the Earth) by the small, somewhat variable angle of about 5 °.

Since the Earth and the Moon move together in space, it must have a somewhat greater speed if it is to orbit the Earth; one has to think of the spatial motion of the Moon as composed of its own and of the Earth.

The mean velocity of the Earth around the Sun is about 4.1 geographic miles or 93660 Parisian Feet in one second; the mean velocity of the Moon in its orbit around the Earth is only 0.13 miles, or 3046 Parisian feet.

The orbit of the Moon, then, does not form a closed curve, because the centre of its motion is also in constant motion, but a wavy line of elliptical nature, and as such it would appear if observed from the Sun.

Suppose for a moment the Earth is at rest, the time would be called the orbital period of the Moon, which will elapse as it passes through the entire circumference of the sky, starting from a fixed point in the sky (a fixed star), until it returns to the same point. This cycle is called the true or sidereal period; it is on average:

= 27 Days 7 Hours 43 Minutes 11.5 Seconds

But since the Earth has its own space movement, it is clear that after the time of a single sidereal orbit the Moon must still travel a certain part of its orbit in order to take that same mutual position in relation to the Sun, which at the beginning of this orbit had taken place. It therefore uses more time for this orbit, which is called synodic period, and that is on average

J. F. J. Schmidt, *The Moon*, Historical & Cultural Astronomy,
https://doi.org/10.1007/978-3-030-37269-9_1

29 Days 12 Hours 44 Minutes 2.8 Seconds

Following this orbit, the lighting-conditions (phases) and our ordinary names of the Moons age are aligned, like the solar and lunar eclipses. [2]

Chapter II
Parallax of the Moon

The angle at which an observer in the centre of the Moon would see the apparent radius of the Earth in the sky is called the parallax of the Moon.

But as the Moon changes its distance from the Earth, the parallax also changes; it increases with decreasing distance of both bodies and vice versa.

Usually, the distance of the Moon is assumed to be the average, at which the parallax is very close to 57 arcminutes.

Since the Earth is not a perfect sphere but is elliptically flattened because of its rotation, one must differentiate between the horizontal equatorial and the horizontal-polar parallax.

This is 1/500 smaller than that.

The distance of the Moon from the Earth changes approximately between 55,000 and 48,000 miles.

The relation between the variability of the parallax and the distance from the Earth can be seen from the following figures; [3]

Parallax	Distance	Parallax	Distance
53' 30"	55225 geogr. Miles	57' 30"	51384 geogr. Miles
54' 00"	54715	58' 00"	50940
54' 30"	54214	58' 30"	50506
55' 00"	53722	59' 00"	50077
55' 30"	53239	59' 30	49657
56' 00"	52762	60' 00"	49243
56' 30"	52294	60' 30"	48837
57' 00"	51834	61' 00"	48437

The parallax of the Moon changes between the values of 62' and 53'.

Their mean value is with a very high degree of certainty = 57'2"32 and this corresponds to a mean distance of the centres from Earth and Moon = 51800 miles. [4]

J. F. J. Schmidt, *The Moon*, Historical & Cultural Astronomy,
https://doi.org/10.1007/978-3-030-37269-9_2

From the above explanation of the parallax of the Moon, it is evident that for an observer on the Moon, the parallax of the Earth must be equal to the angle at which we see the radius of the Moon.

All astronomical data, especially all values of ephemerides relating to the Moon, are for the centre of the Earth, and it is the parallax which, for every place on the surface of the Earth, causes the observed position of the Moon with that of the ephemeris to not match.

The parallax removes the Moon from the zenith, causing it to go down later and earlier, affecting longitude and latitude, Right Ascension and Declination.

But all effect by the parallax changes of the lunar location in the sky can be calculated in great accuracy, when taking into account the flattening of the Earth.

Having carried out these reductions, the observation becomes comparable with the precomputation of the ephemerides, and one can state how closely the observation agrees with the theory. [5]

Depending on the distance, or what is the same thing, on the parallax of the Moon, its apparent size is also known.

The radius of the Moon is assumed to be about 15' 32" at the average distance [6], and its range lies between the limits 14' 41" and 16' 45".

It may be said that the diameter of the Moon in the sky is incidentally half a degree, or that it is equal to a 1/720th part of a great circle in the sky.

Apart from the distance of the centres of both bodies, the Moon is closer to our point of view on the surface of the Earth, the closer it approaches to the zenith.

Between the horizon and the zenith position of the Moon, its distance from the observer may vary by 860 miles, i.e. the distance of the radius of the Earth.

On this occasion, it may be appropriate to recall the well-known illusion by which we often believe that the diameter of the Sun, and especially that of the Moon, is of extraordinary size, when they are very near the horizon.

In reality, there is no optical enlargement, but instead in both cases the direct opposite takes place.

As a result of refraction, all stars appear higher above the horizon than they would be seen without it.

This value on the horizon is about 34'.

Suppose that the lower edge of the Moon touched exactly the line of the horizon, and the diameter of the Moon at that time was really 32', then an observation at that time would not yield the vertical diameter 32', but only about 26'.

The horizontal diameter of the Moon remains unchanged, but the area of the lunar disc is considerably reduced, and it is clear that we are seeing the Moon smaller in size but not, as deception says, larger.

The second cause of the reduction in the diameter of the Moon, as already mentioned, lies in the fact that it increases with increasing altitude, and thus has its relatively lowest value, at the horizon.

This is easily seen from the following figures [7]

If the Moon is 32' 0" in diameter at any time during moonrise, this size is included

0° Height	= 32' 0"	40° Height	= 32' 21"
10° "	= 32' 6"	50° "	= 32' 25"
20° "	= 32' 11"	60° "	= 32' 29"
30° "	= 32' 17"	70° "	= 32' 31"

But since there is evidently no cause to be substantiated on the basis of physics which makes possible an apparent enlargement of the Sun or of the Moon on the horizon, and since a direct measurement on the horizon proves only a diminution, there remains nothing to assume but an illusion of a peculiar kind, whose rather minor differences, however, cannot have been missed by attentive observers.

The sometimes-striking enlargement of the Sun when I have seen it, was always perceived only sparingly, when the area of the horizon, which was very close to it, was richly filled with objects of various kinds and distinct outlines.

The appearance remains the same when dark, sharply defined cloud masses were near the Sun.

But in all cases, when I saw the Sun setting on the shores of the Baltic Sea, the North Sea, and the Mediterranean, and with completely cloudless horizons, on the distant horizon of the sea, it seemed to me no larger than usual.

The proximity of a cloud, or of a ship in the direction of the Sun, immediately recalled the illusion.

Similar things can be observed with the Sun's settings on the horizon of treeless plains.

With the Moon, I find the illusion most conspicuous when, almost at its fullest, it rises to the visible horizon in the dusk.

Its rising in the middle of the night, be it on the horizon of the sea or of the country, this illusion displays to me a little or not at all.

We are constantly led to the explanation, that we involuntarily transfer the exact or approximate size of the objects visible on the horizon simultaneously with the Sun or the Moon, to the diameters the heavenly bodies, and that this translation becomes more and more difficult the more the involuntary comparison of those sensory impressions loses its easiness.

So, if the Sun and Moon rise up against the zenith, we can no longer compare their diameters to trees and houses on the horizon with one and the same glance.

This concept leaves much to be desired, but the truth, whose search might be partly amenable to mathematical treatment, should not be too far removed.

Incidentally, it is not new, but has been noted in a similar form a long time ago. [8]

Chapter III
Size and Mass of the Moon

The Moon sphere is 468½ geographic miles in diameter, 1470½ miles in circumference.

Therefore, it is (according to MÄDLER)

in diameter	= 3.67	times smaller than the Earth.
of surface	= 13.44	
of volume	= 49.25	

The surface of the Moon contains 689240 geogr. Square miles, the physical volume is 53806000 cubic miles.

Therefore, 49¼ Moon spheres would be required to form a sphere of the size of the Earth.

The astronomers also know that the Moon also has a lower average density, according to the investigations of LINDENAU, it is assumed that its mass = 1/88 of the Earth's mass. [9]

This number is to be understood as the weight of 88 Moon spheres corresponding to equalling the weight of the Earth.

It is also known that when the weight of a water ball of the size of the Earth is 1, the weight of the true Earth is 5.44, from which the total weight of the whole Earth mass can easily be calculated to be about 130000 trillion hundredweight.

The 1/88th of the Moon gives the Moon the weight of about 1477 trillion hundredweight: numbers, which in themselves are incomprehensible, as are many others in astronomy, are, by the way, of no scientific value.

If the density of the Earth = 1, then that of the Moon = 0.5614, or the former = 5, assuming the value of the Moon = 2.8. [10]

From this, according to other known laws, the drop height on the surface of the Moon is found in the first second = 2,314 Parisian feet: a gravity 6½ times smaller than that which takes place on the Earth. [11]

J. F. J. Schmidt, *The Moon*, Historical & Cultural Astronomy,
https://doi.org/10.1007/978-3-030-37269-9_3

Chapter IV
Rotation and Libration

At the same time that the Moon completes its true orbit around the Earth, it also turns once around its axis, from which it necessarily follows that it always shows only one and the same side to the Earth.

The Moon's tendency to rotate at a wholly constant angular velocity appears independent of the irregular angular motion of the Moon in its orbit around the Earth, and causes us to perceive noticeable shifts of all the dark surfaces on the surface of our satellite during a lunar cycle.

For example, in the full Moon we have seen a certain spot right in the middle of the Moon's disk, so, when from here the Moon has advanced 90° in its orbit, it will no longer occupy the centre, but have moved away from it by a certain arc, for example to the east.

We will notice a few days later that the patch is again approaching the centre and finally crossing it to move westward away from it.

Similar phenomena are to be found in some distinct grey markings very close to the edge of the Moon; they approach the edge, and leave it again; some disappear to periodically reappear.

According to the immediate impression of the senses, it seems as if the Moon is turned back and forth a little against our eye in the course of a month, so that the patches are soon seen moving to the left, sometimes to the right.

This shift works roughly in the direction from east to west.

A second cause shifts the markings from north to south and vice versa, and this is due to the fact that the lunar orbit is tilted by 5° with respect to the ecliptic, so that when, e.g. the Moon is 5° north of the ecliptic, can look beyond its southern pole into the far-side hemisphere, or in other words, visually all markings increase their apparent distance from the south point of the Moon.

The first shift is called the libration in longitude, the second shift is called the libration in latitude; the constant total effect of both the general libration.

J. F. J. Schmidt, *The Moon*, Historical & Cultural Astronomy,
https://doi.org/10.1007/978-3-030-37269-9_4

Combining the most favourable circumstances, i.e. For example, the maximum of a western libration in longitude coincides with the maximum of the northern latitude, then (according to MÄDLER) a spot lying below the 40th degree of the northern or southern selenographic latitude, the apparent centre of the Moon may approach 10° 24', or so away from it. [12]

So we see these shifts from Earth.

To an observer on the Moon, the globe in the sky would describe slow oscillations about a certain point, which analogously would have to be interpreted as a libration about the average Earth location.

Incidentally, from the explanation of rotation and libration, it is evident that an observer in the apparent centre of the Moon's disk will see the Earth in its zenith, and that for him the Earth will periodically move around the zenith in the course of a month.

All the lunar areas visible to us as outermost edge regions see the Earth rising and falling alternately on their horizon, and beyond that zone the Earth is invisible. — All the irregularities of the lunar motion that we perceive from the Earth are reflected, as it were, in the lesser movements of the Earth around a point that never changes its position for its horizon.

A third type of libration is called the parallactic [*Diurnal*]. Two observers, one at the pole, the other at the equator of the Earth, do not see at the same time the same profile of the lunar disc; while one sees a point in the middle of the lunar disc, the other does not find the same point in the middle at the same time.

The physical libration, very small in its effect, is so far only theoretically known; their numerical value, how large this would seem to be for the Earth, has not yet been sufficiently ascertained by the observations.

Incidentally, it is of no importance to our purpose. [13]

Between the planes of the lunar equator, the lunar orbit and the ecliptic, the following relations take place:

1. The inclination of the lunar equator to the ecliptic is constant = 1½ degrees. [14]
2. The nodes (average points) of the lunar equator coincide with the opposite nodes of the lunar orbit in the ecliptic always completely or extremely close together.
3. The ecliptic always lies between the lunar orbit and the lunar equator. [15]

This information may suffice; they too are actually dispensable for the purpose of this description of the lunar surface, and only for the better understanding of the last section may they serve one who is no longer clearly aware of the bodily position of the Moon and its motion, and who is unable to read other works on it.

Chapter V
Historical Review of the Selenographic Works Over the Last Two Centuries

There can be no particular study of the surface of the Moon before the invention of the telescope (1609).

Moreover, after that time only the efforts of those who observe the mountains of the Moon themselves, i.e. drawn, measured and described; for what in ancient times, and not less in our day, certain authors who were not themselves observers, and for whom often even the elements of science remained unclear, having speculated or created fables about the Moon, every one lacks value, and is for the greater part, after serious refutation, quite worthless.

Then, GALILEO turned his attention to the mountains of the Moon.

He immediately recognized them by their changeable shadow, by the manner in which their summits slowly emerge from the night of the Moon, or go out into it.

He also noted the effects of the Libration and was probably the first who tried to represent the mountains of the Moon in drawings.

Among the astronomers of the seventeenth century, the venerable HEVELIUS of Danzig [Gdansk] has acquired the most merit for a thorough observation of the surface of our satellite, although he had at his disposal only those means which, though exceptional in their time, would today appear to us to be insufficient and most difficult to apply.

He undertook to draw the Moon in its various configurations of light in order to get to know the mountains more closely in form and position, and finally to compose a general map, which may be called the best among the older Moon maps, although it was drafted only by eye. [16]

HEVELIUS gave the names of earthly landscapes and seas to the lunar mountains and the grey plains, which he dubiously considered to be waters. [17]

His contemporary, Father RICCIOLI, who issued a very inadequate map, preferred to abandon the HEVELIUS nomenclature by adding the names of famous men to the lunar mountains, and these names have remained in use to this day. [18]

J. F. J. Schmidt, *The Moon*, Historical & Cultural Astronomy,
https://doi.org/10.1007/978-3-030-37269-9_5

Apart from a few other attempts, mention is made here only of CASSINI's map, which, exceeding the details of HEVELIUS's, was drawn only by eye, and can claim no other than historical interest.

It was not until the middle of the last century that the famous TOBIAS MAYER supplied a very carefully drafted map based on numerous measurements, which for the first time provided a true picture of the mutual position of the lunar mountains. [19]

When in the last quarter of the last century W. HERSCHEL put to use his mighty Telescope with such extraordinary success, the Moon remained almost forgotten.

The great man preferred to explore the depths of the sky, and, against his extensive investigations of the fixed-star world, his rich observations of the bodies of our solar system appear to be of smaller magnitude, despite their far surpassing all that was formerly known.

At the same time, however, SCHRÖTER appeared in Lilienthal near Bremen as an enthusiastic admirer of observational astronomy.

With powerful, for the requirements of his time, excellent mirror telescopes, which he usually manufactured himself, he made the mountains of the Moon the object of many years of investigation; in a great Work, with many illustrations, he published them in 1791 and 1802. [20]

What the value of his works is (and not enough is known to date) and why, on the other hand, SCHRÖTER's constant efforts to discover new changes on the Moon (before a corresponding topographical foundation was created) is less appreciated, and cannot be proven here.

Since the time when SCHRÖTER's work ceased, nothing has happened with the observation of the lunar surface (if one except occasional notes of KÖHLER and KUNOWSKY [21]), until when LOHRMANN of Dresden made the decision to determine the topography of the Moon according to correct mathematical principles.

Provided with good resources, talented, circumspect, precise in his work, he dedicated himself for years to the special observation of the Moon, drew all the clearly recognizable mountains, and surveyed the principal ones for the correct position in the gradation of degrees of his map, without, however, getting involved in height measurements of the mountains, which have already been attempted by HEVELIUS, were practically carried out by SCHRÖTER in many examples.

LOHRMANN divided his map into 25 sections, of which 4 appeared in a special work in 1824. [22]

The joy of being able to hand over his great and splendid work to his contemporaries was not to be.

He suddenly died, when the editorial effort of the work, especially that the engraving of the excellent map was barely half finished

Irrespective of the work of LOHRMANN, independent of all the previous observations of the Moon and possessing those qualities which were praised by LOHRMANN, MÄDLER, with the assistance of W. BEER, began in 1830 his extensive work on the Moon, the results of which are great topographic maps on 4 sheets and a special selenography, published in 1837. [23]

They realised what LOHRMANN was attempting; they have provided the basis that has been lacking since then and created a representation of the Moon with which all earlier ones cannot be compared any longer.

Only LOHRMANN's still unsophisticated works are those who not only endure the comparison, but compete in all relations with MÄDLER'S for the prize of excellence, and remain of the highest value for the latest future of selenography.

LOHRMANN, in the first part of his description of the Moon, confined himself to making known the results of his measurements, and to explaining the mountain landscapes depicted in the first four sections, he avoided any assumption, any daring hypothesis; and this was all the more important, as GRUITHUISEN's fantastic views on life on the Moon had already conjured up an army of hypotheses at that time, some of which still continue to this day, sometimes enjoying a certain fame, and a large part of the Lunar inquisitive public will shift the true position of an unprejudiced serious consideration of nature.

If GRUITHUISEN (Baron Franz von Paula (Franciscus de Paula) Gruithuisen) had long ago overstepped all the limits of the plausible in his imagination, his followers, who, as we see from their writings, lacked any faculty for an independent scientific judgment, went further, and so it has nowadays come to pass that only in Jokes questions about the Moon addressed to the astronomers and thinks to make and maintain opinions, which, even with a superficial criticism, must at once present themselves as a little remote from the absurd and the ridiculous.

— With the same seriousness as LOHRMANN and as a skilled astronomer of subject (which characteristic LOHRMANN could not boast), MÄDLER in his great work has likewise confined himself to a sober description of mountain landscape, and only in a few concluding remarks touches on the field of conjecture, as far as this is concerned justified by obvious analogies.

LOHRMANN and MÄDLER teach us about the mountain forms of the Moon; MÄDLER also occasionally shows what (as far as can be known) is not on the Moon, and cannot be.

Eighteen years have passed since the publication of MÄDLER's work, and since then no new independent work on the mountains of the Moon has become known.

Of the state observatories equipped with all necessary equipment, none ever performed such work.

This must neither be alienating nor provoking censure, considering that the special and long-term study of a single celestial body cannot be the task of astronomical practice, whose entire activity is entirely occupied by initially providing only local determinations, to determine the movement of many stars and to satisfy a highly developed theory.

It is scarcely necessary to recall how brilliant the astronomers have been in the last 18 years; we owe them not only for great theoretical work, but priceless collections of regular observations of the stars, or the discovery of more than 30 planetary objects and a far greater number of comets: a result which, in kind, has no previous epoch for this science.

Also, the consideration of individual planets and the Sun has not been completely neglected; In particular, the multiplicity of the rings of Saturn, the stripes of Jupiter, attempts to understand the spots of the Sun by means of very numerous observations, although it is not to be overlooked that the private observatories here have the most merit. - It took only these brief notes to highlight the merit of the present astronomers, where somebody would be inclined to misjudge them.

The purely scientific part of astronomy differs from the sum of certain facts, which stand as isolated results of observation.

The one who chooses the former, according to his purely theoretical aspirations, feels as much satisfied that he should not be glad to leave the determination of numerical values to change;

and who observe heavenly bodies, i.e. in the astronomical sense, not to look at them only in the telescope, but to determine their place, to measure their motion or their dimensions, is so superfluous because of the magnitude of today's requirements that it is not limited to the frequent consideration of the surface condition of individual celestial bodies, but also to renounce some other, which is otherwise the knowledge to promote and to increase the appeal of life is suitable.

The secondary importance, which is often attributed to the physical observation of the heavens, is also due to the fact that the isolated and partly uncertain perceptions of individual heavenly bodies do not show any definite connection with other disciplines of natural science;

but they will be there one day, and the lacking of our knowledge in this respect, as well as the difficulty of knowing it yet, gives no reason to cease all efforts in this direction, or, judiciously, to contemplate, or even not to continue researching.

When a great and famous astronomer deigns to speak or write even about things that cannot boast of the elegance of a strictly mathematical mode of treatment, it is called, and certainly in some cases not unjustly, flexibility; but whoever deals with physical observations and is not considered of authority, in the case of benevolence, will at least advise the scholars of the so-called Rigourous Theory not to squander time, and to consequentially turn to only promising studies.

Here is, you have to reasonably admit, a lot of truth; but the private view of individuals about what is at stake in the progress of science as a whole, and what value should be attached to certain gaps still to be filled, does not mean that one should not test everything for oneself in another direction of aspiration, and keep the best for oneself.

What theoretical astronomers, who have never observed themselves, by introducing certain analogies into the scholarly and unlearned public, is mostly unimportant and does not affect science at all, but arouses only a temporary interest because of the confusion of ideas that is thereby caused; or because of the credulity which manifests itself in the greatest extent, when incredible and marvelous things are told of the heavenly bodies.

What physicists have occasionally included in their studies from the astronomical fields (in contrast to the earthly phenomena), promises the amazing successes for the future, and there is manifold evidence of the desire to broaden ideas, to raise the point of view of nature and to regard certain phenomena no longer limited to the

Earth, but to conceive them as the effects of universal laws which similar to gravitation, to which heavenly bodies are generally affected.

For a long time, magnetism was considered to be a peculiar phenomenon of the Earth; but since the slightest influence of the Moon and the Sun of the recent years on the three phases of the Earth's magnetic force, the variation, inclination, and intensity had been recognized, attention had to be drawn to the relations which in relation to magnetism despite the long distances between the Earth, its companion and the Sun.

Thus, in the future, the Moon will also provide geologists with a new means of taking a survey of the surface structure of a celestial body, and of comparing the results of thorough research on the lunar mountains with those which are difficult and, in their interpretation, often doubtful Genesis of the Earth have been gradually determined.

For this purpose, the way is still long; Not everyone knows how to appreciate the difficulties which oppose the telescopic contemplation, and especially of a definite interpretation of the forms on the Moon;

a comprehensive geological knowledge is not enough to find the correct explanation at once by the mere sight of a Moon map, and the complete familiarity with all mountain forms of the Moon is not enough to compare them readily with similar forms on Earth.

So everything remains piecework; the geologist is not an astronomer and vice versa, and two centuries of the most energetic scientific activity in all directions have been unable to make reasonably tangible the usefulness which geography may one day derive from the most careful observation of the lunar surface.

Chapter VI
Special Attempts to Represent the Surface of the Moon

In the previous section the efforts of those observers have been discussed, to whom we owe for creating Moon maps, which were drafted from drawings and measurements made at the telescope.

It also remains to recall experiments by which the picture of the Moon was fixed photographically, or by the method of Daguerre, finally to the three-dimensional representations.

When DAGUERRE's great invention became known, it was soon thought that, through the process of producing photographs, we would also capture and locate all the heavenly bodies, although one sometimes indulged in very exaggerated hopes.

It is difficult to see how much we could expect from using the microscope on the daguerreotype, since it had to be known that in proportion as the fixated image is enlarged, so too are the countless amalgam spheres which make the picture visible, and the fine cracks in the polish of the plate increase the fact that beyond a certain limit, one can no longer trace the fineness of the picture.

In fact, not much has been done in this direction; however, the two following daguerrotype of the Moon, of which I know, deserve a special mention.

Of these, one is from the Königsberg observatory, Dr. WICHMANN had the almost full Moon captured by BERKOWSKI by connecting the Daguerre's apparatus along the rotational axis of the 8-foot Heliometer.

This not quite 2-inch-wide image is highly perfect in its form; It shows very subtle differences in the light of the lunar disc, and gives the character of the full Moon much more faithfully than any drawn map can ever represent.

But even the application of a strong loupe reveals the roughness of the plate, and one finds that such a picture will only be of use if it can be made 5 to 7 times larger, at the same brightness and precision. [24]

The other daguerreotype is owned by A.V. HUMBOLDTS; In a picture, not even three inches in size, it shows the crescent-shaped, rising Moon, and one can clearly see from its jagged rows of light the greater inwardly shaded ring-shaped mountains, and sometimes their central mountains.

J. F. J. Schmidt, *The Moon*, Historical & Cultural Astronomy,
https://doi.org/10.1007/978-3-030-37269-9_6

However, the Konigsberg picture is noticeably less pronounced in terms of beauty and brightness. [25]

When there were better maps of the Moon, one might have thought of depicting the mountains of our satellite in a three-dimensional way.

However, one seems to have dared to venture into this enterprise only in more recent times, and one may believe that before the beautifully performed work of the wife of MÄDLER, which imitated the mountainous hemisphere of the Moon in wax, something like this had not been attempted. [26]

Such relief representations are of manifold uses; In a way, they literally sift through an overview of what often had to put together through arduous observations of many years; they help the sensible intuition considerably, and the effect of the shortening on the spherical surface and the shadow of the mountains on the relief can be easily understood.

But even if one carries out some of the more precisely researched lunar mountains on a large scale in relief form, the benefit cannot be misjudged if it concerns a special comparison of the mountain forms of the Moon with those of our Earth.

This view, the careful and thorough pursuit of which may one day acquire a special interest, led me to Bonn in 1849, the talent of the local conservator of natural history museums, Mr. Thomas DICKERT, for addressing such representations.

The first experiments which simulated lunar mountains on gypsum plates of 4 square feet were so satisfying that it was decided to work out the whole visible hemisphere of the Moon on the basis of the MÄDLER maps in relief form, 18 Parisian feet in diameter.

This large and uncommon accomplishment, run under my supervision, has succeeded the arduous and skilful formation completely, after having worked on it without significant interruption for 5 years.

The relief gives the colourful appearance of the full-Moon in a uniform illumination; its hemispherical form allows the imitation of the phases, and if a sharp lateral illumination is employed, the shadow of the mountains develops in such a surprising and splendid manner that one can easily imagine the illusion of observing the landscapes of the Moon as if one views greatly magnified at the telescope. [27]

Chapter VII
Geographic Co-ordinate System (Gradation) of the Lunar Sphere

Similar to the way in which the surface of the Earth or the sphere of the sky is divided by a system of circles in order to obtain a position for locational determinations, the lunar-sphere is given an analogous system to indicate the position of its mountains in terms of latitude and longitude.

The degree network of the Earth is identical with that of the sky; one can understand the latter as if the Earth resting in the centre of the heavens had described, by virtue of its rotation, the network of degrees in the sky; by infinitely extending its spin, it designates the two poles of the world; an infinitely elongated radius of its equator describes the celestial equator under the stars, and likewise the infinitely extended radii of arbitrary latitudes describe the parallel circles of the sky.

The first meridian on Earth is arbitrary, e.g. that of Ferro, by which is to be understood the one which stands 20° 0' 0 "west of the meridian of the Paris Observatory.

The first meridian of the sky is determined by the mean point of the ecliptic with the equator (in the vernal equinox); it is changeable because of precession.

With the Moon, the determination of a gradation network is fraught with difficulties, in that all the observations made above appear to be affected by all the irregularities of the motion of the Moon.

It is necessary to observe by effect of parallax, and strictly taken but free from the effect of all four causes of libration, or at any rate from the influence of ordinary libration.

If this happens, then, in view of certain points of the Moon's disk, we find the middle meridian, which bisects the side facing us from north to south; we also find the equator, which bisects the lunar disc in the direction from east to west.

If both arcs; equator and first meridian, appear as straight lines as seen from the Earth, the Moon is in the position of its middle libration. This position is chosen for the map drawing in which the co-ordinate system is designed according to the orthographic projection.

J. F. J. Schmidt, *The Moon*, Historical & Cultural Astronomy,
https://doi.org/10.1007/978-3-030-37269-9_7

Such a map thus refers to the sight of the Moon from an infinite distance, and all the shortenings in the mountain shapes, which depend on the curvature of the globe, are rendered very closely as viewed from the Earth in the telescope.

From the first middle meridian one counts now the longitudes up to 90° to the west as positive and designates them by '+', in contrast to the longitudes counted up to 90° to the east, to which one puts the sign '–', and one calls negative longitudes.

If the Moon is in the south in the sky, then, when we look at it, the half of its disk to the right is indeed directed to the west, the left to the east.

On this basis, it should not be strange to say that the increasing illumination of the lunar landscapes, i.e. the rising of the Sun, begins in the west and that it advances eastwards.

In the last section of this book, which talks of the day and the night on the Moon I did not consider the mentioned use in selenography, but, according to our earthly view, the west is called that region on the horizon of the Moon, where the Sun shines, in the east, where it rises up, for an observer, who in spirit moves to the side of the Moon facing us.

From the equator, one expects the degrees of latitude 90° to the north positive and 90° to the south negative.

One is thus able to designate a mountain according to its longitude and latitude on the Moon, just as on Earth, or as a star in the sky;

Chiefly, four selenographers have been busy determining a certain number of the principal points on the Moon in longitude and latitude.

TOBIAS MAYER was the first; after him LAMBERT made such provisions. [28]

Through the extensive work of LOHRMANN [29] and MÄDLER [30] numerous points on the Moon have been determined with accuracy.

Some examples will give you an idea.

The most accurate location on the Moon is that of the small craters Mosling A, whose eastern longitude WICHMANN has determined from 50 observations = 5° 13' 23", whose southern latitude = 3° 10' 55".

The error seen from the Earth is estimated by WICHMANN to be 1/2 arcseconds, but the error seen from the centre of the Moon or the selenocentric error is estimated to be 50". [31]

Following on from this accuracy, the location of the Central mountains in crater Manilius according to TOBIAS MAYER, NICOLLET and BOUVARD.

Locality of the crater Manilius:

Longitude 9° 02' 06 West. Latitude 14° 34' 01 North; after 27 Observations by Tobias MAYER [32]

Longitude 8° 47' 00 West. Latitude 14° 27' 01 North; after 50 Observations by NICOLLET [33]

Longitude 8° 46' 09 West. Latitude 14° 26' 08 North; after 124 Observations by BOUVARD.

Among the 27 measurements of MAYER, the deviation of the individual observations from the mean, as far as the maxima, is in longitude +31' and –40', in latitude +21' and –17'.

In any case, apart from their probably greater accuracy, the observations of the French astronomers deserve to be preferred because in their time the elements of the orbit of the Moon were already better known.

If we compare the measurements of MAYER with those of LOHRMANN, we have for example:

Location of the crater Dionysius:

by MAYER:					by LOHRMANN:				
1	west. longitude	17° 22'	north. latitude	3° 09'	1	west. longitude	17° 02' 2	north. latitude	2° 54' 8
2		17° 11'		2° 41'	2		17° 29' 3		2° 55' 5
3		17° 54'		2° 38'	3		17° 32' 2		2° 54' 3
4		17° 14'		3° 07'	4		17° 36' 2		3° 01' 6
5		17° 14'		2° 59'	5		16° 56' 6		2° 47' 1
6		17° 17'		2° 54'	6		16° 46' 8		2° 38' 8
7		16° 50'		2° 46'	7		18° 00' 0		2° 58' 2
8		17° 03'		3° 11'	8		16° 54' 0		2° 37' 1
9		17° 28'		2° 47'					
	Average	17° 17' 0		2° 54' 7		Average	17° 08' 7		2° 50' 9

Biggest deviations from the mean:

from MAYER in longitude: +11' and -27'; in latitude: +16' and –14'
from LOHRMANN in longitude: +51' and -21'; in latitude: +11' and –14'

If we examine the individual results of LOHRMANN and MÄDLER in more detail, we find that in longitude as well as in latitude deviations of ½°, on both sides of the mean, are very common, but it is not to be overlooked that this magnitude is from the Earth, appears only at a very small angle. [34]

At the same time, it is noted that MAYER's observations, taken individually, equal the precision of LOHRMANN's and MÄDLER'S measurements, which proves once again the brilliant talent of this unforgettable astronomer.

Only the average numbers of his observations do not have that certainty because of the telescopic equipment commonly used 400 years ago, which one may ascribe to those of the selenographers of our time.

There is no doubt that today most of the geographically and nautically important points on our Earth are more accurately defined in longitude and latitude than the main points on the Moon.

The location of some observatories, i.e. their distance from the equator and the difference in longitude from the meridian of Ferro is so accurately known that, especially with regard to latitude, the error still to be determined is only to be sought in the decimals of the second.

The geographical location of many cities, mountains and islands is known within one minute of arc.

Of a large number of other points on very remote coasts, which are visited very little, or where the unfavourable circumstances and the less carefulness of the observers have had an injurious effect, considerable uncertainty may be presupposed.

Half a century ago A.V. HUMBOLDT faulted the position of the famous Mexico City near 2°, as determined by earlier geographers. Many longitudes and latitudes of the navigators, which are not based on direct astronomical observations, but derived only from the casually closed chronometer course and the ship's calculation, are, in particular, differing in longitude, sometimes by ½° from each other.

This, at least at the beginning of this century, still took place in some cases.

It is safe to assume, however, that much improved navigation over the past fifty years, and the more rigorous use of nautical observational techniques, will improve much of the past errors. [35]

Chapter VIII
Causes, Which at Different Times Make the Moon mountains Appear Altered

Apart from the optical changes which the lunar mountains undergo, in their position and shape, by the shortening on the spherical surface, but then by the effects of libration, there are others, which affect, to a much greater degree, the external appearance of individual mountains or entire landscapes, or appear to be completely temporarily transformed. This effect is due to the changing illumination of the mountains by the Sun.

For all lunar landscapes, which we see during the waxing phase directly on the terminator, the day has just begun; for their horizon, the rising of the solar disk has taken place, in whole or in part. [36] From the terminator of the waning phase on the Moon you can see the setting of the Sun. According to their height, then the mountains cast more or less long shadows, whose black, often sharp-edged figures afford a very beautiful sight. [37] All around it the tallest ridge of the already lighted wall shimmers as a narrow gold fringe, and often, like a star, the summit of a central mountain rises from the darkness of the depths, which the light of the Sun had just struck. [38] As the Sun rises, the scene changes; the shadows on the mountains and in the craters shorten; their gradual disappearance brings forth objects that were previously covered by them. [Compare this to the frontispiece and the adjacent drawing, which represent two interesting shadow covered landscapes of the Moon.]

With the disappearance of the last trace of shadow, the extraordinary sharpness of the outline is lost, in which, at the time of lower Sun altitude, one can recognize the horizontal dimensions of the mountains, and soon it is scarcely possible to observe the highest illumination in the same area, i.e. at the time of full Moon, to find well-known forms that used to stand out with their colossal proportions. [39]

For every lunar landscape, there is twice a maximum of the clearest visibility in each lunation, which, apart from the effects of libration, depends entirely on a certain height of the Sun above the horizons there.

For very shallow areas, which are occupied only by insignificant hills, one must, in order to observe reliably, determine the time of sunrise or sunset;

for a high mountain region and for the countless craters, where the dark shadows obscure so much, the time to investigate is to choose in which the Sun changes its altitude between 3° and 11°.

J. F. J. Schmidt, *The Moon*, Historical & Cultural Astronomy,
https://doi.org/10.1007/978-3-030-37269-9_8

If one follows the appearance of a certain landscape from day to day, one notices the slow change, until the characteristic colouring of the full Moon gradually appears, and one is then able to identify dominant points of light and streaks with localities, looking in completely different shape for the eye during the lunar phases.

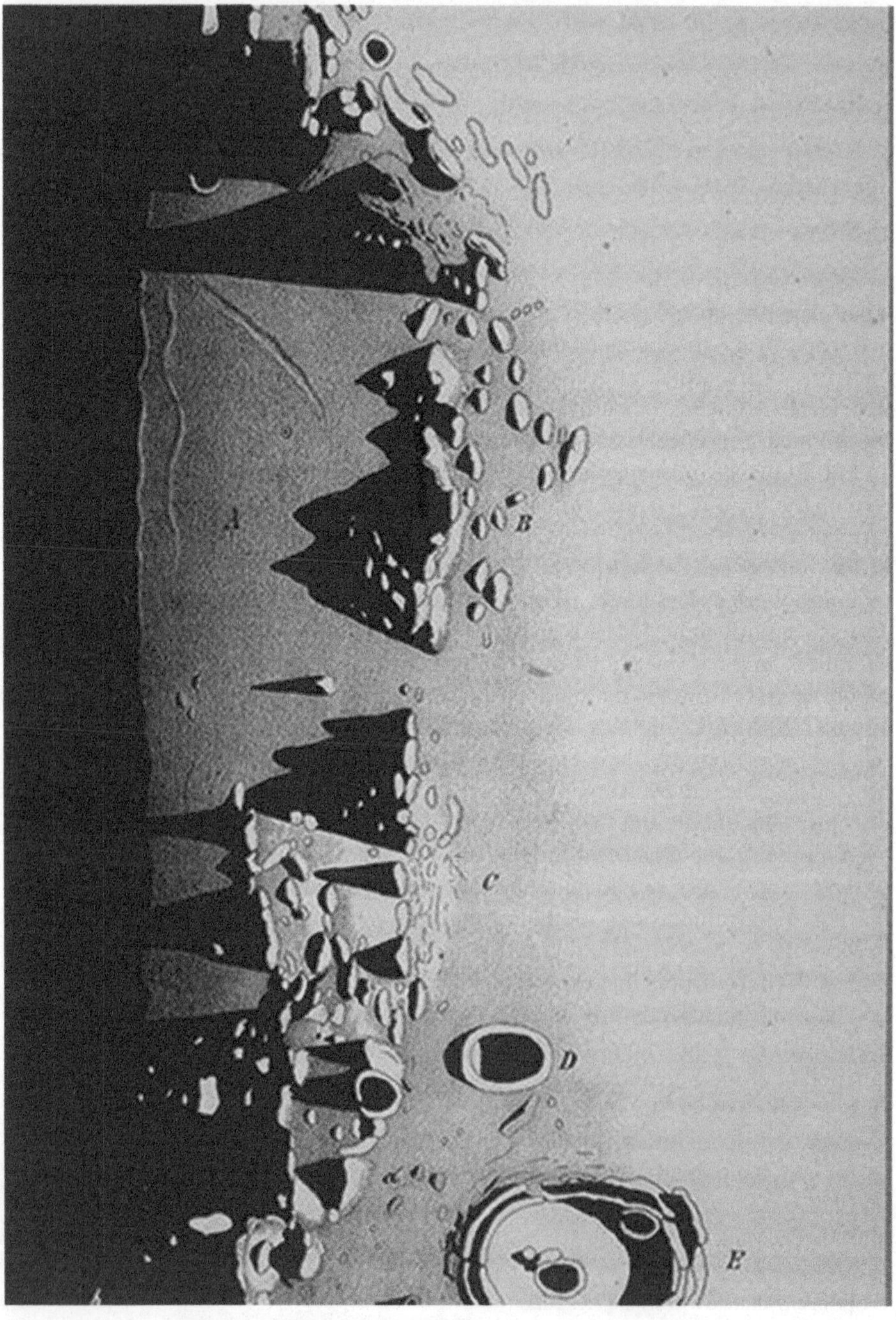

Shadowcast of the Moon mountains Caucasus at sunset

The full Moon shows only differences of light and colour; the main impression is afforded by the large dark areas already visible to the unaided eye; these first the radiations of narrow streaks of light: examples of which are spread all around Copernicus and Tycho.

But the endless detail, the thousands of mountains, hills, and little craters that often astonish the observer during the phases, have vanished without a trace to the greater part.

Because of the lack of shadow they remain unnoticed, because, moreover, they are neither sufficiently distinguished from their environment by brightness nor by peculiar colouring.

It can be seen from what has been said that one cannot draw a topographical chart at the time of full Moon; for every single Moon region, one has to wait, when the Sun has only a small altitude above its horizon.

Many years are required, especially in the astronomer's unfavourable climate of central and northern Europe, until, even with a telescope of only moderate optical power, one can exhaustively explore and draw a small area.

Chapter IX
About the Mountain Shadows in Particular

The shadows of the Moon mountains are completely black; their contrast to bright mountainous terrain is often so great that the eye of the observer working with the Micrometer is greatly strained using powerful telescopes, that one may believe that similar extremes do not exist on Earth.

On the Moon, because of the lack of an atmosphere, light reflected sideways or even radiant from the heavens, cannot penetrate into the shadows, and the only possibility, especially in the deep craters, is provided if that a powerful reflection from a wall opposite the Sun and already illuminated by it, diminishes the darkness of the shadow a little.

However, this is beyond us without specific practical knowledge.

The shadow remains deep black, and there is never the slightest trace of the landscape covered by it.

If a long shadow touches the phase with its tip, the clarity at the extreme end is not particularly great; it differs too little from the dim lighting of its surroundings, especially when it lies in the grey plains.

The more the Sun rises, the sharply contrasted appear to show slowly shortening form.

In rare cases, at the edge of the shadow, in deep craters, one sees a faded brownish-grey fringe, which makes the measurement a little more difficult; close by, in a very similar crater, one seeks the apparition in vain. [40]

To explain this, it is very inappropriate to resort to penumbral shadow or to an atmosphere.

It is much more probable that the shadow-throwing crater wall is covered with a multitude of small elevations, or with close-packed blocks of colossal dimensions, which from our distance are invisible to us, their narrow shadows parallel to each other on the opposite wall whose bright interstices we cannot notice, and the result will be that we notice only a washed-out edge instead of the often distinction in the tips of the columns.

J. F. J. Schmidt, *The Moon*, Historical & Cultural Astronomy,
https://doi.org/10.1007/978-3-030-37269-9_9

The same would be the case if such a crater wall were often perforated at its upper wall.

The mountains of the Earth have individual cases of this kind on a small scale, though not frequently. [41]

The mathematical penumbra, which, surrounding the main shadow, is based upon the apparent diameter of the Sun, enlarges on the lunar surface with a sinking Sun; [42] the same thing happens on Earth, and everyone knows that in the evening and in the morning earthly shadows at their edges become indistinct, though for multiple physical reasons this indistinctness is greater than it would originally be only as consequence of the Sun's diameter.

What impression, viewed from great heights, the shadows of our mountains afford, whether in the distance a penumbra is noticeable, whether in the shadow one sees clearly all the objects, and how the same behaves in darkness towards the brightly lit landscape - these are all questions that have so far been answered inadequately. As numerous as the descriptions of mountain journeys in the Cordilleras, in the Himalayas or in the Swiss Alps are, so often in the report on aerostatic ascension the sight of the Earth from a great height everywhere one misses a certain statement as to what contrast the shadows in the landscape cause against their surroundings. [43]

In regard to the atmosphere one would like to believe that, moreover, an observer at 3000 toisen high, precisely because of the effect of the atmosphere, would see the shadows neither very dark, nor particularly sharply delineated.

Because of the lack of other observations, I am permitted to share some small examples of my own experience, which are, of course, far from making any definite decision.

But you may remember that even the everyday appearance, the ordinary shadow, is not without interest.

Looking at the time of sunrise on the summit of the Drachenfelsen in Siebengebirge, the shadow of this famous bell-shaped trachyte mass extends westwards far beyond the left bank of the Rhine, so that, even in the brightest cloudless sky, little is visible on the heavily vegetated and therefore dark soil, but it clearly shows the objects in the landscape.

If one compares (always from the summit) a little later the shadow of the Drachenfelsen, or at another time that of the opposite basalt head of Rolandseck, when both lie on the reflection of the stream, they appear very striking and moderately dark, one is unaware of the penumbral shadow, or diminishing the ordinary clarity of visibility on the random passing ships they cover.

The straight-line distances are respectively 350 and 150 toisen.

In the Carpathians, from the peaks of the mighty mountains Radost, Smrk and Lissahora, their shadows are seen in the morning and evening in only very moderate darkness; in particular, the outline of the former is difficult to recognize when, as is usually the case, it covers the immense dark pine groves of the western valleys and mountains.

The terrain is nowhere bright enough to favour a strong emergence of the mountain shadow.

Only when it is apparently brought closer to the eye, on clouds of mist that are hovering, does it clearly develop its dark shape, and one convinces oneself that only in the vicinity is the extent of the penumbras noticeable at the edges.

If one chooses a favourable point of view in the winter on the plain, or else in the high mountains, one sees in the shadow of the mountains on the shining snow-surfaces a far more perfect phenomenon.

The shadows are seen in a darkness and sharpness reminiscent, albeit very roughly, of the shadows of the lunar mountains.

However great the contrast between light and dark may be in a heavily illuminated snowy landscape, one always, unless the eye has been otherwise blinded by the reflection, one always sees in the shadows of a not very high mountain, on whose summit one stands, all objects clearly in form and colour.

On a perfectly serene morning at 973 toisen above sea level on the Wengernalp, one is very close to the much-praised panorama of the Bernese High Alps, and the time waits for the shadows of the monstrous peaks of the Jungfrau and the Mönchs to rctreat over the Trümmleten-Thal, one is astonished at the sharpness of the shadow protuberance, which slowly reaches the glistening snow surfaces, on the snow-covered glaciers.

From distances of, say, 1000 to 2000 toisen, I could not perceive any trace of the penumbra under these circumstances.

In the shadow itself, on the other hand, the smallest objects remained visible: each narrow furrow formed by sliding masses of snow, every break in the ice, and the blue-green lustre of it no less than the distinct colouring of the Felshörner, bare and sharp here and there, breaking the blanket of the everlasting snow.

During the approximately three-hourly observation, the view was never downwards, but always directed upwards against the steeply inclined northern slope of the ice. [44]

If one compares with this perception the statements of those travellers who had the courage to climb the more than 2000 toisen high dangerous snow peaks of the Jungfrau, the Oberaarhorn and the Finsteraarhorn, [45] you will notice a striking difference.

These travellers, though their reports are never directly related to the mountain shadows, unanimously tell that, from above, and in a perfectly clear sky, they had not only a very limited distant visibility, but even in the neighbouring valleys, they had almost nothing in detail clearly recognized, while down below, and even from 18 to 20 hours away at the Jura, the snow mountains of Canton Bern could be seen completely clearly.

So, if the observers at the summit were astonished to see so little detail even in neighbouring valleys, then it is most likely to be true first of your sun-lit land, and hence of the shaded places to a much greater degree, the latter at that time could not have a significant extent anywhere.

This digression has the sole purpose of indicating that, if there were certain experiences of the general manifestations of terrestrial mountain shadows from great heights, it would be easier to judge whether the extraordinary blackness of the mountain shadows on the Moon is only a consequence of the lack of atmosphere there or whether that darkness and sharpness is to be regarded as a consequence of the distance, and of the considerable contrast to bright mountain landscapes.

If one agrees that the darkness of the shadows on the Moon is a perfect one, effected by the lack of an atmosphere, it is clear that an observer standing on the shadow-throwing crater wall will see nothing in the dark depths, even though the Sun is already shining at a height of 7 ° –10 °; it will at most outline weakly illuminated contours through the reflection of the opposite wall.

An observer in the depths, while he sees that bright radiant wall above him, will be surrounded by night, will see above him the sky not in blue but in black colour and many stars, although the day has been unbroken for more than 30 hours.

Chapter X
Earthshine (*Lumen Secundarium*)

When the Moon is crescent-shaped, one sees, with an unaided eye, the remainder of its disk, not touched by the rays of the Sun, in a dull grey light, as long as it is bright enough to distinguish itself from the heavens; when the dawn and dusk have already become very weak, or completely extinguished.

This ash-coloured light is scarcely noticeable in the vicinity of the bright crescent, but better on the opposite edge, and one perceives on such occasion how great the difference of the impression of both light-sources is on our eye, for each time it seems as if the lit crescent does not belong to the remainder of the disc, but as if its radius of curvature were considerably larger.

The same illusion arises in lunar eclipses shortly before or after totality, or generally in any major partial eclipse, and remains visible even in telescopes with very low magnification but great light-gathering power.

It disappears immediately, if one uses the ordinary power of astronomical refractors, and one convinces without any measurement, that the sharply illuminated part of the Moon and the dull shining part form a complete circular line. [46]

The reason for this illusion lies in the so-called overflow of light: a part of the phenomena of irradiation, by virtue of which we see the diameters of the stars larger than they are, and by which certain Micrometre measurements lose their value, if one fails to take into account the effect of irradiation in order to determine the final result.

The weak ashen-coloured light, the cause of which had already been recognized by LEONARDO DA VINCI, [47] GALILEO, [48] and MÖSTLIN [*Michael Maestlin*], is the sunlight, which is reflected from the Earth on the night side of the Moon.

It is easy to see that at the time of the new Moon, when we do not see the Moon because of its proximity to the Sun, and because it turns its darkened half towards us, an observer on the terrestrial hemisphere of the satellite would see the fully-lightened Earth in its sky.

J. F. J. Schmidt, *The Moon*, Historical & Cultural Astronomy,
https://doi.org/10.1007/978-3-030-37269-9_10

The more the Moon grows, the weaker appears its ash-coloured light, because it is more and more overshadowed by the radiance of the crescent, and, moreover, the amount of light reflected from the Earth diminishes day by day.

For as the Moon grows from our view, the Earth seems to be diminishing for an observer on the Moon.

New Moon, first quarter, full Moon and last quarter correspond to our companion in terms of its hemisphere: full Earth, last quarter, new Earth and first quarter, if one wants to keep the analogous designations.

In good telescopes the ash-coloured light appears very clearly, but perhaps not always in the same colour, which is difficult to specify at all.

That this is generally grey, cannot be mistaken; one usually notices that it has a yellow or brown-green addition. [49]

In this light, the larger bright and dark spots are easily found under favourable circumstances, and it remains visible at the telescope much longer than to the naked eye.

Without a telescope, I recognized it on the day prior to first quarter, but with a 3-foot refractor on the third evening after first quarter. [50]

SCHRÖTER's opinion that the intensity of the Earth's light on the Moon depends on whether the oceanic, or the predominantly continental, side of the Earth reflects their light is certainly well founded, but the examination of these and many other questions is not to be expected from the observer under the Central and Northern European skies.

The great brilliance of some points in the night side, such as that of the brightest crater Aristarchus, formerly gave rise to suspicion of burning volcanoes on the Moon, either because one misunderstood an expression of HERSCHEL, or saw what one wished to see.

More than 60 years have not yet been enough to completely eradicate this volcanic hypothesis, even though SCHRÖTER had already found the right explanation, and throughout the time to MÄDLER there was never a lack of observers who were reminded that in every lunation, not only Aristarchus, but also other bright points on the night side can be seen, which are particularly noticeable on the full Moon by their brilliance. [51]

Meanwhile, the observation of the *Lumen Secundarium* should not be neglected.

There may be light phenomena on the dark side of the Moon, unrelated to burning volcanoes, and which, if we should see them, would certainly be worthy of a very careful investigation.

The fact that one sometimes sees a faint telescopic shooting star or, like SCHRÖTER, numerous fragments of a fire meteorite burst in the distance, in front of the dark night side, is, understandably, unconnected with the earthshine of the Moon.

The ash-coloured half of the Moon is seen best in the evening in the spring, or in the morning in the autumn, 3 to 5 days after and before the new Moon. In very favourable circumstances, they can be seen on the horizon with the naked eye, if the bright crescent Moon either has already gone, or just begins to rise above the horizon. [52]

Chapter XI
Apparitions During a Lunar Eclipse

If the Earth did not have the light-disturbing and opacifying atmosphere, and if the Sun were only a luminous point, the edge of the shadow of the Earth, as it travels on the Moon, would appear to us to be a sharply delimited near-circular line.

Observation teaches that at the time of a lunar eclipse, the edge of the shadow of the Earth is very blurred and indistinct, that the shadow itself is not black and covering everything, but more or less red and transparent.

The indistinctness of the shadow profile is due, in part, to the penumbra of the Sun, which depends on the apparent diameter of the Sun, and partly to the atmosphere of the Earth, which is usually filled with fumes and clouds.

Previously, the times were recorded when single well-bounded Moon patches were covered or uncovered by the Earth's shadow.

Such observed at two observatories after mid-local time entries or exits then led to the determination of the difference in longitude; but because of the great uncertainty of observation, this method has long since been abandoned, and to determine the geographic longitudes it is better to use the solar eclipses, star occultations, and for some years the electro-telegraphic signals.

In the case of any partial or total lunar eclipses, it is found that both the general duration thereof and the duration of the eclipsing of individual patches are considerably greater than the prediction indicated.

From observation the eclipse begins earlier, and stops later.

The average of the shadow cone of the Earth where the Moon passes was thus greater than the calculation had shown.

Already TOBIAS MAYER and LAMBERT endeavoured to calculate more precisely the amount of this enlargement of the shadow of the Earth (the magnification coefficients), but this investigation was first carried out in a consistent manner by MÄDLER.

J. F. J. Schmidt, *The Moon*, Historical & Cultural Astronomy,
https://doi.org/10.1007/978-3-030-37269-9_11

The enlargement of the shadow of the Earth is expressed in terms of the shadow radius, as such was the basis of the calculation. Calling this 'A', MÄDLER [53] found

1833 Dec. 26	A = 1/65
1835 Jun. 10	A = 1/28
1837 Oct. 13	A = 1/64
1844 Nov. 24	A = 1/49

From my own observations and from the data of other astronomers I found the following results: [54]

1842 Jan. 26	A = 1/50
1844 May 31	A = 1/56
1844 Nov. 24	A = 1/52
1848 Mar. 19	A = 1/45
1849 Mar. 8	A = 1/44

It can be seen that the magnitude of the enlargement is very variable, that its variations, so far as we know, lie between the limits 1/28 and 1/65.

The accuracy of the value A depends not only on the observation, but also on the reliability of the Lunar Tables.

It must now be assumed that the cause of the enlargement of the shadow of the Earth is twofold, that the value of A is due to the theoretical penumbral shadow of the Earth, the effect of the refraction and the independent shadow of the Earth's atmosphere on the whole extent of our planet, even at the time of the lunar eclipse, the rays of the Sun, which just reach the Moon, form tangents.

Only in very rare cases does the totally darkened Moon disappear, and it cannot be found even in the telescope; but in all other cases it remains visible to the unaided eye, often very bright, and more or less vivid red, the excellent transparency of which permits the perception of very fine points and streaks on the full Moon. [55]

Against the centre of the shadow of the Earth the darkness increases considerably.

The periphery (the penumbra) is light brownish, very washed out and in great darkness, especially shortly before the beginning and immediately after the end of the totality, is surrounded by a very beautiful sky-blue colour.

There is no doubt that the refraction of the Earth's atmosphere is the chief cause of this phenomenon, as has been suggested in olden times, but a sufficient explanation of all variations of colour has not yet been found.

The red of the eclipsed Moon is partly copper-coloured, partly clear, similar to that of the glowing iron, and sometimes so bright that during the totality a halo can form around the Moon, when faint clouds pass by. [56]

When, at the dawn of January 7, 1852, the Moon, completely darkened, was very close to the western horizon of Bonn, the redness diminished itself completely with increasing daylight.

The whole disc just turned greyish yellow; all large areas (the maria) remained visible to the unaided eye, and the sight in the telescope reminded vividly of the effect of the *lumen secundarium*.

A great lunar eclipse is one of the most beautiful phenomena in the sky; since it has been recognized that it is unsuitable for purely astronomical purposes, it has been neglected irrespective of the considerable physical interest.

MÄDLER's generous endeavours still stand almost alone.

But there is no doubt that the study of all the phenomena of a lunar eclipse could be of great importance for the study of the conditions of our atmosphere on the whole.

Chapter XII
Phenomena of the Moon During a Solar Eclipse

If the Moon obscures a part of the Sun, we see its edge, which, depending on the state of the libration, is sometimes filled with high, sometimes low mountains, in perfect sharpness.

Its disc as seen through smoked glass is completely black, but seen with the naked eye, but for the brightness of the sky in the area where the Sun stands.

But at the time of the magnificent spectacle of a total solar eclipse, phenomena occur that have not yet found a satisfying explanation.

If, in my opinion and experience, they are for the most part not in direct relation to the Moon itself, they may be briefly mentioned here for the sake of completeness.

Already more than 200 years ago, it had been seen that the black Moon, as long as it completely obscured the Sun, was surrounded by a bright nimbus, which gradually grew outward; one has much later, in this nimbus and on the edge of the Moon itself, bright red prominences of different shape and size had been observed.

In the total solar eclipse of July 7, 1842, one first learnt of these and other phenomena better.

One saw them again in the total eclipses of the later events, which are so rare for the individual places of the Earth, as: 1850 Aug. 8, especially in the famous phenomenon of the 1851 July 28, further 1852 Dec. 10 and 1853 Nov. 30 [57]

The white radiant nimbus, the corona, used to be thought of as an effect of the lunar atmosphere, without thinking to investigate the refraction of such a huge atmosphere during the darkness on the measured Sun diameter.

Prejudice, which wants to ascribe the air to the Moon at all costs, and other unreasonable opinions, has preserved this view until today in some respects.

Although these coronae now form part of the Sun's photosphere, or may largely owe their origin to the refraction (inflexion) of light on the edge of the Moon, it must be admitted that both explanations are applicable simultaneously and to varying degrees, but in general we still know very little about it.

J. F. J. Schmidt, *The Moon*, Historical & Cultural Astronomy,
https://doi.org/10.1007/978-3-030-37269-9_12

The ruby-red, strangely shaped prominences (protuberances), which I myself observed on July 28, 1851, on an excellent telescope at Rastenburg in East Prussia, constitute the greatest and most remarkable phenomenon of the whole phenomenon.

Without being able to elaborate on the still very differing views on it, I limit myself to the remark that the protuberances, according to my full conviction, drawn directly from observation, belong entirely to the Sun, no matter how ingenious experiments of individual physicists visually may have shown an approximate resemblance to the figures in question.

By virtue of its motion, the Moon covered the prominences to the east and made them appear to grow westward at the edge of the Sun for the same reason, until, after a period of three minutes, all the splendour of the spectacle vanished when the first sunbeam once again glared.

Chapter XIII
Opinions About the Atmosphere of the Moon

It is just as difficult as unpleasant to have to renegotiate on a much-discussed subject, which can be considered as having been long since completed.

This is therefore to be done only insofar as I have revealed here, by consideration of all those hypotheses, phantasies, and fables, which competent and incompetent writers have revealed about it, to simply communicate the clear and decisive statement of BESSEL's, to which some facts will be later added for explanation.

BESSEL says in a treatise concerning the lunar atmosphere the following: [58]

"It has long been known that the light of a fixed star at the moment when it touches the edge of the Moon is not noticeably diverted from its rectilinear motion.

One recognizes this from the comparison of the two values of the Moon's radius, which can be deduced on the one hand from direct measurement and, on the other hand, derived from the duration of the obscuration in front of a fixed star.

If there were a refraction of the rays about the Moon, the second determination would have to make the radius smaller by twice that of the first; whereas, on the other hand, both provisions are so close that one cannot find any decisive difference.

Thus, to attribute an atmosphere to the Moon, one must either assume it to be incapable of disturbing the rays, or one must assert that the edge of the Moon is covered with mountains and the height is so considerable that the stars, by striking the Earth, the edges of it disappear and reappear, are seen through an already so thin layer of the atmosphere, that they no longer suffer any noticeable refraction of rays there.

The first assumption seems to curtail all further discussion, but it is far too much in breach of the analogy of all the fluids we know, to be correct. The second was really done by the defenders of the lunar atmosphere.

However, they have failed to compare the density of the atmosphere at the height of the supposed ridge-mountains and the surface of the Moon by Mariotte's law, and therefore, without being able to justify their ideas, believed that, in spite of the deficiency, or the Irrelevance of the refraction of the rays in the height of the assumed mountains, the atmosphere on the surface of the Moon itself, a considerable density, which SCHRÖTER estimates to be 1/29 of the density of our air."

J. F. J. Schmidt, *The Moon*, Historical & Cultural Astronomy,
https://doi.org/10.1007/978-3-030-37269-9_13

BESSEL then shows that between the two values of the lunar radii, which is derived on the one hand from solar eclipses and star occultations, and on the other hand from meridian observations, only by 0"13 (in degrees of arc) are different from each other, which size is much smaller than that which is still to be concerned and inevitable mistakes of the results.

He then shows that the very high ridge mountains of 4" altitude belong to the exceptions and that only in a few particular cases a star touches such mountains, but will often enter and leave the middle region of the Moon, where, as a rule, ridge mountains of only 2" are not very often seen.

BESSEL, then, in favour of the defenders of a Moon atmosphere, uses the most favourable and greatest value in his calculations; he finds that the assumed Moon air cannot have more than 1/968 of the density of the Earth's atmosphere.

Even if pure oxygen gas were substituted for the lunar air, the density would be only 1/863 under the specific conditions.

Here the temperature is set to 0°; but in order to find the density = 1/500 for the lunar air consisting of pure oxygen, the temperature should be 240 degrees of the hundred-part thermometer below the freezing point.

-BESSEL concludes with the following words:

"It follows from these developments that the compatibility of irrelevant or totally absent refraction at the lunar edge with a considerable density of the atmosphere on the true surface of the lunar body cannot be as readily admitted as the defenders of the lunar resemblances believed on the Earth.

In fact, I know no means of preserving the Moon an atmosphere, as the assumption of a non-refracting one.

Whether one prefers to do this, or to seek new explanations of the observations, from which one has deduced the presence of the atmosphere, I have to wait ....

I believe that after considering these doubts against the explanations of perceptions taken for hints of a lunar atmosphere, there will be no reason for the existence of them which would be more probable than the reason that the imperceptibility of the refraction of rays at the edge of the Moon gives them.

To fully explain the fortitude of this reason was the purpose of this essay, which, incidentally, is not written for those who find pleasure in dreaming of similarities between the Moon and the Earth: in this I do not wish to bother them with theoretical reasoning."

— BESSEL.

The study of the famous astronomer, whose statement has been given here, thus leads to the result that the phenomena of the star obscuration and solar eclipses contradict the assumption of a lunar atmosphere.

It has sometimes been remarked that the disappearance of a star on the edge of the Moon is not so sudden, and as far as the instantaneous loss of light is concerned, it was not done in the usual way.

But it is not far-off, the cause of it rather in momentary changes of our atmosphere, and what is far more likely to be sought in rapidly transient modifications of our eyesight strained at such observations.

Among perhaps 100 stellar occultations I have observed, very few cases have appeared to show any deviation from the ordinary course, and any unbiased observer will have had very similar experiences without thinking of lunar air.

If, therefore, we are compelled to deny the Moon not only of one of our similar but also every atmosphere, it is still necessary to recall certain local phenomena which were first brought to our attention by MÄDLER, and which, even if they did not create an atmosphere in any one form may seem to imply, but may indicate that nature is still active in different forms on the Moon.

In rare cases close to the phase around single peaks illuminated by the Sun, one notices a blue light to a small extent, but mostly at very bright points, without discovering anything similar on neighbouring mountains.

MÄDLER for example saw the apparition with Archimedes, Aristillus, and Autolycus; I and others on the other hand saw it on the west bank of Clavius, the Maurolycus and at the western foot of the mountain Pico, while it lay exactly on the terminator.

Meanwhile, in these observations, the highest caution is indispensable, since even the first-rate achromatic telescope has brilliant ring-shaped mountains, such as Aristardi and Tycho, and also other bright objects, e.g. white stars of the first magnitude, are seen surrounded by blue light.

Is the mentioned phenomenon real, and can we all find that in a few cases a mountain shadow appears semi-transparent, or that a deep crater can be seen shadow less at the phase, all that remains is to assume local exhalations of three dimensional vapours expelled with high velocity, covering cloudlike areas of the lunar disc for us. But we are still a long way from a sure decision in this regard.

For us then, the Moon remains a body without trace of an atmosphere, which can be proved by convincing observations, after finding that no refraction, no diminution of the lunar edges, no atmospheric turbidity and precipitation as clouds and snow on the Moon, as on the Earth, is yet to be discovered.

-As a result of this lack of atmosphere, the observer enjoys the unclouded sight of the lunar disk, whereas Earth's surrounding atmosphere itself does prevent observation unfortunately all too often.

He sees the mountains near the edges of the Moon in the same clarity and brightness as those in the centre; he finds no trace of mists and clouds, and never sees the Polar Regions or the peaks of the mountains on the Moon covered with snow.

Chapter XIV
Regarding the Surface of the Moon

The extraordinary wealth of the formations, or, to avoid here a geological expression [59] of the forms on the Moon, makes a coherent and yet reasonably clear description very difficult.

What is always advisable, is to maintain the Classification of the Moon mountains, introduced by SCHRÖTER, LOHRHANN, and MÄDLER, so one does not fail to feel the variability in the systematic sequencing of manifestations, which are more or less related to one another by infinitely different intermediate and transitional elements.

The composition, as well as horizontal and vertical dimensions, are what will occupy us mainly in the following sections; the descriptive selenography has to do only with this object of observation; one can engage in further assumptions, but one also leaves the solid ground of a science whose first foundations are to be found in numerical determinations.

Thus, while for the time being we are only concerned with the sure part of the observation of the Moon, it is necessary first to give the means by which one obtains knowledge of the height of the mountains.

The height ratios on the Moon become of greater importance the more one approaches the objective, which is to provide an explanation of the origin of the lunar mountains and the special comparison of these with the mountains of the Earth.

The method of height measurements and the certainty of their results, proven by special examples, must precede the description of the mountains.

J. F. J. Schmidt, *The Moon*, Historical & Cultural Astronomy,
https://doi.org/10.1007/978-3-030-37269-9_14

Chapter XV
Height Measurements

On Earth we determine either the absolute height of a mountain, its elevation, or its relative elevation over points of its surroundings. We measure with the mercury barometer, with the thermo-barometer and with the aneroid barometer, and where the locality does not permit this, by triangulation.

Finally, if necessary, it remains possible, through use of shadows, or through the point of the furthest visibility of a mountain on the sea, to establish its height. By the three instruments mentioned above, when we climb with them from the plain to a mountain, we measure the decrease of the atmospheric pressure, and from the amount of this reduction we find how many toisen we have increased. If a mountain is unclimbable, its height must be determined by measuring the angle.

At a moderate distance from the mountain, one orients a horizontal line whose length is exactly known in toisen. From the endpoints of these, one measures the apparent elevation or the zenith distance of the summit, corrects them for earthly refraction, and calculates the height of the summit above the line by simple formulas, by measuring the distance of the summit from the endpoints of the line that has been determined.

If the height of the latter above the sea level is also known by a direct levelling, or by barometric observation, then the absolute height of the mountain can be obtained.

Nobody will want to try the last two of the above-mentioned methods, since they are inherently very inaccurate for physical reasons, and in any case, presuppose precise instruments anyway.

If one observes at a certain time the shadow of a mountain, i.e. you notice the point of the shadow tip on the ground, and determine by another operation the exact distance of this point from the true base of the mountain,

(By which is meant that which lies vertically under the summit and at the same height of that plain, whereupon the shadow was observed) thus, a slight astronomical calculation first gives the apparent solar altitude, and then, by applying trigonometry, one finds the summit of the mountain above the plain.

J. F. J. Schmidt, *The Moon*, Historical & Cultural Astronomy,
https://doi.org/10.1007/978-3-030-37269-9_15

This method, which is very inferior to terrestrial mountain measurement, is to be applied to the lunar mountains with some alteration and somewhat greater difficulty, if one wants to achieve the most reliable results possible.

There are, however, two other ways to approximate the height of the mountains of the Moon.

These and the 'Method of the Shadows' may be briefly considered.

XV.1 Method of Light Tangents

GALILEO and HEVELIUS were the first to attempt this theoretically correct but practically barely usable measurement technique.

They waited, for example when, at the time of the last quarter (i.e. when the *terminator* forms a straight line), a mountain peak in the night side of the Moon lost the last ray of light from the Sun.

At this moment, they measure the distance of the summit from the terminator.

The last ray still striking the mountain touches the terminator, is its tangent, and forms with it an angle of 90° on the lunar disc.

Now, according to the measurement, is it known that the distance of the mountain from the phase, and also the radius of the Moon, both e.g. expressed in toisen, the simple trigonometric calculation gives the height of the mountain sought.

Because, however, the first light and the last light of such a summit cannot be perceived with certainty for a variety of reasons, and because the assumption is made that the last ray of sunshine is a tangent to the middle surface of the Moon, which is very seldom the case, one can ascribe this method only a small usefulness. It has only rarely led to approximated results.

Thus, HEVELIUS found a peak in the Apennines of the Moon = 2643 toisen, about 200 toisen too small, and SCHRÖTER thus determined the height of the mountain Hadley to be 1969 toisen, which he had calculated from several shadow measurements on average to be 2093 toisen.

XV.2 Profile Measurement

If a mountain lies exactly in the edge of the Moon, so that we have a real side view of it, nothing would be simpler than to determine its apparent elevation with the micrometer at such time,

if the size to be measured is not very small even in the most favourable case and the decision as to whether the foot of the mountain is really in the extreme would be quite impossible.

Through the libration the apparent position of the mountain is constantly changed; soon it lies in the other hemisphere of the Moon, and is invisible, now it lies on the limb, and again at another time the libration on this side puts it on the side of the Moon facing the Earth.

For this it is always very difficult to obtain a certain judgment about the middle lunar limb, especially at the South Pole. SCHRÖTER has dealt with these measurements, or rather estimates, often.

Such observations by MÄDLER are as follows:

Ridge mountains at the South Pole: [60]

1832 July 12	Height	= 4184 Toisen.
1832 „ 13	„	= 3617 „

In this case, the measurement itself is assumed to be quite correct, the change effected by the libration from one day to the next = 567 toises. In a similar way I measured a southern edge mountain as follows:

1854 February 13	Height	= 3576 Toisen
29 hours later		= 2406 „
Difference		= 1170 Toisen

but the second measurement is very uncertain. I measure a very high dome shaped South Pole mountain:

1854 October 8	Height	= 4223 Toisen; and 27 days later:
1854 November 4	„	= 4286 „

We must be content with the subject of Ridge mountains, knowing that they are very high, perhaps not inferior to the peaks of our Himalayas. A great certainty of their height determinations is not to be expected.

XV.3 Method of Shadows [61]

The safest way to get to know the heights of the Moon mountains is to measure the mountain shadow and the distance of the mountain from the terminator:

Everything else required for the calculation can easily be found with the aid of the astronomical tables.

The two values to be measured are in many cases subject to more or less optical shortening; one measures their projections, and adjusts them by the calculation of the influence of the shortening.

By the known apparent distance of the Moon from the Sun at the time of the observation one finds the size of the phase of the Moon, by this and by the distance of the mountain from the phase, the apparent height of the Sun on the horizon of the mountain, and finally the length of shadow expressed in degrees of the arc of the Moon, the height of the Sun at the end of the shadow.

From the ratio of these two Sun heights the relative mountain height is determined.

It is a relative one because the calculation gives only the height difference between the top of the mountain and that point on the surface of the Moon which was struck at the time of observation by the endpoint of the shadow.

If an isolated mountain lies in the open plain, a far greater accuracy may be achieved;

but if the shadow falls on the plain, sometimes to a hundred or two hundred toisen, or to the lower point of the land, it is indispensable to indicate at what altitude a given measurement is made, for the length of the shadow, consequently making its way over the mountain and valley depends on the changing height of the Sun on the horizon of the measured summit.

The citing of the Sun's height each time is also necessary in the event that the mountain has no sharp point, but flat or dome-shaped rounded summit. If the latter takes place, its shadow will indicate the true height only at the smallest possible Sun height, but in all other cases too low a height is measured. As the Sun has reached 1 and 2° of altitude, the mountain, no matter how high it may be, can no longer cast a shadow.

However, for a cone-shaped mountain located in an open plain, the inclusion of the elevation angle of the Sun at the time of measurement is less necessary. If such a mountain is for example 1200 toisen high, and if one finds values that are at one time 900 toisen and other times 1500 toisen, then for good reason one must assume considerable errors of observation, or must assume that, regardless of the conical shape of the mountain, its summit is rounded. The last circumstance can then be determined by measurements at high and low Sun altitude.

If one measures the shadow of a crater wall at quite different times, one may not immediately conclude from observation values differing by several hundred toisen, that the crater measurement is inaccurate, for if the crater forms a basin-shaped depression, one must extend the lower wall heights, the more the shadow edge moves away from the centre of the crater. Here the need for specifying the height of the Sun is most important.

It goes without saying that you have to distinguish between the east wall and west wall, even in the seemingly regular ring mountains. For example, on the important crater Cuvier, as MÄDLER correctly noted from his measurements, the eastern wall raises 500 toisen higher above the base than the western one.

After these remarks, it will be of interest to use various examples to ascertain the degree of certainty that we may attribute to the mountain heights on the Moon.

But as it might seem doubtful to some people to expect a particular accuracy from surveys made at distances of 50,000 miles, it is not improper to examine in advance, as it is today with the knowledge of the mountain heights of our Earth. We shall find that, with the exception of a few cases, there are still significant errors to be feared for many mountain heights of our Earth. The causes of this are well known; the barometer measurement at both stations may be very accurate, but the hypothesis about the heat dissipation may be irregular.

Of course, among the Tropics, where the changes in air pressure and heat are so regular and so small, these errors are to be expected less, but nevertheless, at high altitudes, differences between the figures of two observers of twenty toisen are not uncommon.

The trigonometric measurement of a mountain can be performed with high perfection, but the confidence of the end result (i.e. the altitude) depends, on the one hand, on the variable refraction, and on the accuracy with which the elevation of the base can be determined.

As a first example I choose one of the most famous mountains on Earth, the Pic de Teide [Mount Teide] of Tenerife. It has often been climbed and measured many times since the last century.

According to the information given by two works (by HUMBOLDT and v. DEVILLES) [62], I put the results known to me together here, without, for the time being, making a choice in terms of accuracy.

Date	Observer	Height of Mount Teide	Measurement method
1724	FEUILLE	2213 Toisen	Trigonometric (from the sea)
1724	VERGUIN	2025	Barometric
1742	HERNANDEZ	2658	Barometric
1749	MANKVIILETTE	2000	Trigonometric (from the sea)
1752	HEBERDEN	2408	Barometric
1771	PINGRÉ & BORDA [63]	1742	Barometric
1771	PINGRÉ & BORDA	1701	Trigonometric (from the sea)
1776	BORDA	1905	Trigonometric
1776	BORDA [64]	1976	Barometric
1785	LAMANON	1902	Barometric
1788	CHURRUCA	2193	Trigonometric (from the sea)
?	JOHNSTONE	1899	Trigonometric (from the sea)
1803	CORDIER	1920	Barometric
1816	L. v. BUCH	1868	Barometric
1837	DUMOULIN	1901	Barometric
1842	DEVILLE	1901	Barometric

If one did not wish to make a critical selection here, but consider the exceptional deviations of the results among themselves as observational errors, one would be able to say that the difference between the largest and smallest quotations is 957 toisen or 5742 Parisian feet, more than half the true amount.

It is clear; however, how even a superficial glance at the above figures will show that, according to the nature of the methods used, it is necessary to make a choice.

This selection first fails to consider any heights observed at sea level, for reasons which are easy to overlook, if one remembers that the distances of the ship from the center of the mountain are based only on rough estimates from the ship's log.

It is also permissible for good reasons, which are not discussed here, to exclude the old observations of 1724.

Doing so, we still have the following measurements of the pics of Tenerife:

in Year	1776 Height =	1905 Toisen	Trigonometric	from BORDA.
" "	1770 "	1976 "	barometric	from BORDA.
" "	1785 "	1902 "	barometric	from LAMANON.
" "	1803 "	1920 "	barometric	from CORDIER.
" "	1816 "	1868 "	barometric	from L. v. BUCH.
" "	1837 "	1901 "	barometric	from DUMOULIN.
" "	1842 "	1901 "	barometric	from DEVILLE.

Average = 1910.4 Toisen = 11462 Parisian feet.

The deviations of the individual data from the mean are thus as follows, if the mean = 1910 toisen, ignoring the decimal point:

+ 5	Toisen.
– 66	"
+ 8	"
– 10	"
+ 42	"
+ 9	"
+ 9	"

Hence, if every observation is ascribed to the same degree of accuracy, a mean error of any value of +/– 33 toisen or 1/58 of the whole height, and a probable error of the arithmetic mean = +/– 8 1/3 toisen, or 1/240 of the total height.

Without the two barometer measurements of BORDA and L. v. BUCH, the average would be 5 measurements = 1906 toisen, the mean error of each determination = +/– 8 toisen and the probable error of the mean = +/– 2 3/5 toisen = 1/794 of the whole.

This accuracy is very satisfactory; Therefore, not taking into account the barometric observations of 1776 and 1816, because they do not agree with a value assumed to be close to correct, is quite inadmissible as long as one is unable to justify such an exclusion on reasonable critical grounds.

So if we look at the penultimate result, we can assert with a certain approximation to the truth that the height of the Pics of Tenerife is well known except for an uncertainty of 1/240 of the whole, provided that the height of the formerly active, now long-extinct volcano does not change slowly.

I would cite as a second example the elevation determinations of a smaller volcano, Vesuvius, which have been repeated more frequently since 1749. [65] If it were not almost certain that the mountain would slowly change its height, may this change only affect the crater edges, or concerning the eruption cone forming at eruptions.

As for the height of the Chimborazo, I can only cite the three following partly barometric and partly trigonometric measurements: barometric, inasmuch as a part of the triangulation was to be completed by a barometric leveling; because never before has it been possible to carry an instrument to the top of this mountain. [66]

Height of Chimborazo =	3220 Toisen from	LA CONDAMINE and BOUGUER.
" " "	3380 " "	JÖRGE JUAN.
" " "	3350 " "	A. v. HUMBOLDT.

That the latter result is the surest should not be doubtful. The greatest accuracy, however, has so far been attained, as it seems, in determining the altitude of Mont Blanc, Monte Rosa, and Kintchinjinga [Kangchenjunga] in Himalaya. [67] After having learned to appreciate the degree of accuracy of the mountain measurements on the Earth with a few examples, we turn to the mountain measurements on the Moon, but only the method of the shadows is to be understood.

We first consider a more or less steep mountain, which at least for the most part can cast its shadow on flat land.

I. The Moon hill *Calippus a*, on the northeast edge of the Mare Serenitatis, highest peak of the Caucasus.

At the time of the rising Moon (first quarter), this high peak, which slopes down to the east at an angle of about 35°, casts its shadow 21 geographical miles to the plain area of the Cassini ring mountain.

The westward retreating shadow never meets very uneven land; only at the foot of the mountain itself it lies on a steeply rising ground.

It has been measured 1 time by SCHRÖTER, 3 times by MÄDLER and 24 times by me.

In all the following examples, *H* represents the elevation angle of the center of the Sun on the horizon of the mountain, *h* the relative height of the mountain expressed in toisen.

The composition of the values begins with the lowest position of the Sun; it ends when the mountain shadow has become too small for the measurement when the Sun is too high.

I designate the measurements of SCHRÖTER by (SR.), Those of MÄDLER by (M).

My own observations follow without a label.

It is clear that one will find too small a mountain height if one measures the shadow at a time when it does not yet appear fully developed, or in other words when it is still bluntly cut off from the terminator.

Only at the moment when the tip of the shadow separates from the terminator does one obtain the whole height of the mountain. This is easy to understand from the following numbers.

I found the examples. (Fractions of minutes are omitted):

1854	Dec. 26.	With the Sun altitude	H = 4° 12'	Mountain height	h = 2406 Toisen.
	Mar. 6.	" " "	H = 4° 15'	"	h = 2455 "
	Dec. 26.	" " "	H = 4° 19'	"	h = 2534 "
	Dec. 26.	" " "	H = 4° 21'	"	h = 2567 "

In all these observations, the shadow flowed together with the overall night of the Moon, and its tip could not yet be seen. On the other hand it was measured: 1853 June 13th at the height of the Sun H = 5° 9' mountain height h = 2951 toisen. Here the shadow point was already clearly visible, and the calculation also resulted very close to the already known otherwise highest height of this summit.

The complete measurements of the mountain (*Calippus a* are the following: [68]

When H =	5° 09'	h = 2951 Toisen	when H =	6° 31'	h = 2262 Toisen
	5° 11'	2920		8° 26'	2255
	5° 13'	2979		8° 34'	2198
	5° 23'	2752		8° 36'	2301
	5° 23'	3190 (M.)		9° 11'	2257
	5° 29'	2840		11° 04'	1721
	5° 48'	2741 (M.)		11° 31'	2307
	6° 03'	2656		11° 49'	2496
	6° 03'	2638 (M.)		13° 09'	2332
	6° 04'	2633		13° 23'	1615
	6° 15'	2706		14° 33'	2143
	6° 27'	2622 (SR.)		21° 34'	2189

The first thing that stands out about these numbers is the decrease in mountain height as the elevation of the Sun increases; one finds a difference of 1575 toisen between the measurements at H = 5° 23' and H = 13° 23'.

It must now be considered that the measurement at 131/2° solar altitude is already very uncertain because of the brevity of the shadow, and that a small observation error has a large impact on the mountain height.

Furthermore, evidence from the locality suggests that at 131/2° solar altitude the shadow falls on considerably higher terraces already adjacent to the mountains, while there is sufficient reason to believe that the mountain forms a very rugged peak to the east, and therefore could cast shadow at a solar elevation from 13° to 14°.

If, however, one were to regard the deviations of the above numbers merely as observational errors without considering the nature of the land on which the shadow is cast, it would be necessary to admit that any measurement, however good, would be affected by your average error of close to +/– 400 toisen, an assumption which is so completely rejected by all experiences that it need not be pursued further.

Obviously, it is necessary to combine the measurements made at nearly equal elevations of the Sun in arithmetic means in order to compensate for the observational errors and to make them less detrimental. If this is done, then:

With H =	5° 11'	h =	2900 toisen	from	3 observations	
	5° 20'		2973		3	=17838 Parisian Feet;
	5° 23'		2971		2	
	5° 31'		2881		4	
	5° 47'		2745		3	
	6° 03'		2642		3	
	6° 24'		2530		3	
	8° 32'		2251		3	
	9° 11'		2257		1	
	11° 28'		2175		3	
	13° 42'		2030		3	= 12180 Par. Feet

In this way the observations without coercion gain a different form; the very regular decrease in the altitude is observed with increasing altitude, and the observation to the contrary makes the assumption that the gradual elevation of the ground towards the mountain is much greater than the rounding of the summit.

The apparent diminution of the measured altitude of 943 toisen between a sunrise of 5° 20' and 13° 42' is due to the fact that the shadow retreats from the plain to the eastern slope of the mountain, to a lesser extent, however, from the fact that at more than 13° solar altitude the top, perhaps even dome-shaped vaulted summit could cast no shadow

If one considers the deviations of the second survey to depend solely on the unevenness of the terrain, one finds the mild error of a measurement only = +/– 112 toisen = 1/26, this uncertainty is still somewhat lower than that in the three-dimensional elevation of the Chimborazo. However, I am of the opinion that the uncertainty of +/– 112 toisen is much too great for *Calippus a* (if one excludes measurements at more than 9° Sun altitude).

It is certainly not more than +/– 50 toisen in very good observations; but the conclusion will only be possible if 200 to 300 observations are available!

II. Cap. Huygens [*Mons Huygens*] in the Apennines [*Montes Apenninus*], after 20 observations by myself, and 4 observations from MÄDLER.

Combining all observations using arithmetic means, we have:

With H =	4° 40'	h =	2495 toisen	from	4 observations
	5° 25'		2389		4
	6° 20'		2488		4
	7° 34'		2434		4
	8° 31'		2389		4
	11° 03'		2195		4

If the uncertainty of H = 11° is disregarded, the extremes are 106 toisen, or 1/25 of the whole, assuming that the shadow always falls into the plain and the summit is pointed. Strictly speaking, however, both do not happen.

III. Alpen η. After an observation of MÄDLER and 8 of mine, one has:

With H =	3° 58'	h =	1396 toisen	from	4 observations
	5° 21'		1326		4
	7° 33'		1150		3

The shadow falls eastwards into the very flat Mare Imbrium.

But the figures demonstrate that the mountain has a round crest. The same is true of many of the other summits, which form this marvellous mountain range, which is the most glorious in the rising Sun.

IV. Cap. Laplace, after a great deal of my own measurements.[69]

With H =	3° 33'	h =	1674 toisen	from	2 observations
	3° 37'		1701		2
	3° 44'		1653		2
	3° 51'		1679		2
	3° 59'		1666		2
	4° 12'		1611		2
	5° 03'		1725		2
	6° 10'		1738		2
	6° 38'		1668		2

Average = 1679 Toisen = 10074 Par. Feet

The difference between the largest and smallest figures is = 127 toisen; Deviations from average = – 68 and + 59 toisen. If the mean deviation from the arithmetic mean = 63 toisen, this is about 1/27 of the whole height.

The shadow always lies in the very flat area of the Mare Imbrium, but its tip is so extraordinarily narrow and sharp that I have never been able to detect with complete certainty its separation from the phase.

The true mountain height is certainly at least 1700 toisen, almost exactly equal to the height of Mount Etna.

For all this information must be distinguished: the mean and probable error of a result, determined by the calculation; the greatest extremes between the lowest and highest values; the deviation of one or more observations from an arithmetic mean.

For nine mountains I have included a part of these measurements in arithmetic means derived from observations at very low angles of the Sun, and calculated the probable error of these mean numbers, assuming that the deviations of the measurements are due to observation errors only. I found:

For	*Cap Huygens*	probable error =	1/155 of the whole height
"	*Huygens*		1/117
"	*Pico*		1/63
"	*Lahire [Mons La Hire]*		1/87
"	*Pico A*		1/58
"	*Alpen* η [Montes Alpes]		1/51
"	*Calippus* α		1/49
"	*Eudoxus* α		1/93
"	*Eratosthenes* η		1/34

These results are very satisfactory considering all the difficulty of the observation and the small number of measurements; These results demonstrate to us that we know the height of some Moon-mountains with a certainty, for which in mountains of our Earth in many cases at present we do not yet have.

So far, there has only been talk of isolated mountains whose shadows can never fall on quite uneven ground.

Now let us look at the cases so exceedingly frequent on the Moon, in which the wall of a deep crater can cast its shadow, usually into the bottom of the concavity, now against the steep sides of the wall.

As an example may be chosen the highest mountain of the Moon (*Curtius d*), which is still determinable by the shadow, which lies on the northeastern rampart of the 111/2 mile wide crater *Curtius*, and with a waning Moon its shadow is cast in the northern part of the crater.

In this western direction of the shadow, MÄDLER once measured the mountain.

From my 40 observations of this extraordinary mountain, I select those which are useful for our purpose.

with H =	22° 00'	h =	1100 Toisen	from	2 Observations
	20° 12'		1412 „		1
	12° 21'		3106 „		4
	11° 33'		3761 „		4
	10° 28'		4046 „		4
	9° 44'		3936 „		3
	8° 34'		3834 „		3
	7° 33'		3770 „		3
	6° 16'		3261 „		2
	5° 38'		2994 „		4
	5° 22'		2903 „		3
	2° 46'		1048 „		1

These measurements are to be explained as follows.

If the Sun is still 22° high above the horizon of the mountain when the Moon is waning, then its shadow, which is very short at that time, gives only a relative elevation of the summit of 1100 toisen; The shadow barely reaches the inner highest wall of the crater.

When the Sun has sunk to 13°, the shadow already covers all the eastern terraces, and begins to expand westward on the recessed crater floor; one finds here the height = 2800 toisen.

At 101/2° solar altitude, the shadow tip touches the centre, that is the northern part of the meridian of the middle of the crater, for because of its position the shadow can never reach the true centre of the crater.

Then the relative height difference is probably at least 4200 toisen or 25200 Par. Feet.

From here, the bottom of the crater begins to rise again against the western wall, and the numbers clearly show how the height of the mountain is less and less represented by the calculation when the Sun sets.

At 2° 46' the tip of the shadow is on top of the north-west crater wall, which is over 2000 toisen, and soon it merges into the overall night of the Moon. [70]

Similarly, for the western wall of the *Tycho* crater:

With H	= 141/2°	h =	1940 Toisen,
	= 101/4°		2208 „
	= 8°		2350 „

At a height of 14°, the shadow has scarcely concealed the western interior terraces; At 8°, on the other hand, it extends to the middle, up to the base of the central peak, which is about 1000 toisen high.

The crater wall, which throws its shadow into the caldera as the Moon rises, must lose the shadow of the outer environment as the Moon wanes. It is clear that the measurements made at both times must decide whether the mountains really are a cauldron-shaped; below the height of the surface of the Moon, or not.

For this, of course, no measurement is required, for while the interior of the craters is often filled with black night shadows for days, with illumination opposite, one sees little or no shadow developed outward on the same wall; It's easy to see for yourself that many craters are true circular holes that barely have a sizeable raised edge around them.

It will, however, be interesting to take a closer look at this subject, which is important for the genesis of the lunar surface, and to give the results, which emerged from the measurements of it.

It suffices to give only round numbers, for the most part, that neither MÄDLERS nor my own doubly numerous observations in their union suffice to set the differences in height even in the most favorable cases to +/– 50 toisen.

I show in what follows the height difference of the crater's edge against the lowest point in the crater, i.e. the inner relative height, by J; but the height of the edge above the external environment of the crater through A.

Only those ring-mountains are chosen, which lie either in a smooth plain, or in regions where small irregularities in the terrain are of little importance.

The depth (J-A) will therefore be the depression of the craters below the average surface of the Moon [*'Sea level'*], while, as already mentioned, J itself denotes the depth of the crater base beneath the base of the surrounding mountains

Picard	J = 980t A = 420 J − A = 560		Geminus	J = 1970t A = 700 J − A = 1270
Picard *A*	J = 1160t A = 400 J − A = 760		Plinius	J = 1350t A = 550 J − A = 800
Picard *d*	J = 1200t A = 300 J − A = 900		Eudoxus	J = 1140t A = 500 J − A = 640
Burckhardt	J = 2300t A = 700 J − A = 1600		Aristoteles	J = 1370t A = 400 J − A = 970
Aristillus	J = 1450t A = 900 J − A = 500		Aristarchus	J = 1100t A = 400 J − A = 700

Autolycus	J = 1420t A = 800 J − A = 620		Bullialdus	J = 1400t A = 790 J − A = 610
Triesnecker	J = 850t A = 350 J − A = 500		Werner	J = 1900t A = 760 J − A = 1140
Copernicus	J = 1840t A = 670 J − A = 1170		Abulfeda	J = 1400t A = 400 J − A = 1000
Reinhold	J = 1350t A = 380 J − A = 970		Theophilus	J = 2000t A = 470 J − A = 1530
Landsberg	J = 1250t A = 430 J − A = 820		Piccolomini	J = 2000t A = 600 J − A = 1400
Timocharis	J = 1100t A = 550 J − A = 550		Hommel C	J = 1280t A = 600 J − A = 680
Lambert	J = 900t A = 380 J − A = 520		Jacobi	J = 1480t A = 500 J − A = 980

These 24 examples are selected from a large number of other observations.

In order to be as brief as possible, I do not state whether the eastern or western crater wall is used, which of MÄDLER's observations are used, and why I have occasionally used numbers - which do not quite agree with MÄDLER's.

Let us examine how often the outer wall A is contained in its height in the inner wall height J, so if we ask for the ratio of the opposite heights of the crater mountains, we obtain the following numbers for it:

Picard	= 2.3	Picard *d*	= 4.0
Picard *A*	= 2.9	Burkhardt [*Burckhardt*]	= 3.3
Geminus	= 2.6	Timocharis	= 2.0
Plinius	= 2.4	Lambert	= 2.3
Eudoxus	= 1.8	Aristarchus	= 2.8
Aristoteles	= 3.4	Bulliald [*Bullialdus*]	= 1.8
Aristillus [*Aristyllus*]	= 2.6	Werner	= 2.5
Triesnecker	= 1.7	Theophilus	= 4.2
Copernicus	= 2.5	Piccolomini	= 3.3
Reinhold	= 3.6	Hommel *C*	= 2.1
Landsberg [*Lansberg*]	= 2.9	Jacobi	= 2.9

The average of all these numbers is 2.73, that is the inner drop of a crater mountain is about twenty-three times as great as that of the outer, and it is therefore irrefutable that the Moon crater is deeply submerged below the surface of the Moon. If we consider the great difficulty of measuring the shadow of the crater mountains, the uncertainty as to whether we always had the same point of the wall, whether we really have reached the maxima of the two heights, I require at least 20-30 observations for each individual crater in order to ascertain the ratio of the external and the internal elevation with certainty; but we are still very far from it.

If these numbers are arranged according to the localities in which the crater occurs, the ratios found are:

2.6 for craters	in the dark terrain,
2.7 „ „	in bright terrain,
2.8 „ „	in hilly surroundings,
2,8 „ „	of considerable size and depth, which also contain terraces and central mountains.

It is not without surprise that the independence of this proportion is related to the great differences of the terrain, from which the crater has most probably broken through some sort of eruption.

It is still necessary, as long as we are concerned with the measurements of the mountains, to recollect the central mountains which are isolated, or, in small groups, often designate the center of the crater.

In particular, how the increasing and decreasing shadow of a mountain range serves to level the depth of the ground to a certain extent by means of measurement,so you can also apply the measurement of the shadow of a central mountains,by means of which it is possible to determine the curvature by which the crater floor is formed, but which is always incomplete, because the central summit with its deep position, is very frequently covered by the shadow of the walls, and always assuming that its summit is a steep peak.

We have for example:

Northern Central mountain in Theophilus:

When H =	8° 36'	h =	1123	Toisen
	5° 39'		1055	„
	4° 53'		925	„

i.e. at 81/2° the peak of the shadow is still in the depth of the crater between the base of the central mountain and the lowest rampart; at 5° 39' it crosses the lower terrace, and at a height of 4° 53' it exceeds the next higher.

Further, the measurement cannot be continued, for soon the Central mountain disappears in the shadow of the massive crater rim, which is about 2,000 toisen high.

If one differentiates between the middle inner wall height J, and between the height of a mountain towering over the wall over the depth = J', so it is, although only very approximately possible, the ratio of the height of the Central mountain to the internal (relative) wall height J.

The following figures serve to illustrate this proportion, which are useful to express how often the height of the central mountain range is contained in the height of the surrounding mountains:

Cleomedes	= 2.3	Gassendi	= 1.7
Copernicus	= 4.6	Bettinus	= 3.0
Pythagoras	= 2.6	Moretus	= 2.0
Tycho	= 2.2	Ricciolus	= 2.7
Walter	= 1.7	Lilius	= 2.2 [71]
Arzachel	= 1.9	Albategnius	= 2.4
Werner	= 2.8	Maurolycus	= 4.0 (?)
Theophilus	= 1.8	Piccolomini	= 1.9
Cyrillus	= 2.2	Pitiscus	= 2.4
Langrenus	= 1.7	Ulacq	= 2.2
Petavius	= 1.9		

From the average, these 21 examples show that the central mountains are 2.4 times lower than the mean valleys above the depth of the crater, and since it was previously found that the ratio of the external to the inner elevation is 1:2.7, it follows that the central mountains, with their summit, do not yet reach the mean level of the surface of the Moon: a result which MÄDLER first drew attention to.

This is illustrated by the following examples, where I choose the ring-mountains, in which the external and internal elevation, as well as the elevation of the central mountain range, is sufficiently known.

Of course, one does not have J, but (J-A) i.e. the absolute depth of the craters below the surface of the Moon

Copernicus J-A	= 1170t	Central peak =	370t	Difference =	–700t
Werner ,,	= 1140	,,	700	,,	–440
Theophilus ,,	= 1520	,,	1200	,,	–330
Piccolomini “	= 1400	,,	1000	,,	–400

Accordingly, these 4 central mountains are so deep that their peaks project an average of 400 to 500 toisen down from the surface of the Moon: a remarkable proportion, which is well deserved to be examined more closely in many other examples.

We have seen that in favorable cases the height of a moonlit mountain can be determined very accurately, and it is considered that only by these height measurements alone are we able to truly shape individual mountains and, more importantly, the shapes of the crater Basins after studying the purely topographical work, these investigations will have to be regarded as one of the main tasks of selenography.

In geography one notes with great care all changes that affect the surface of our planet, especially by volcanic forces.

It is these changes which stimulate scientific inquiry as well as human imagination.

For the geologist, there is nothing solid and changeable on Earth;

perhaps almost every hour their surface shakes forth there; mountains are created and others collapse.

Over the last three centuries these have been ascended:

the Monte Nuovo near Naples and the Mexican Yorullo; [72] it is collapsed:

the summit of the Carguairazo and earlier the summit of Capac-Urcu:

both neighbours of the Chimborazo, the former separated from it by a valley, the other one opposite to the east in the eastern Cordillera. [73]

If such changes are possible even on Earth, regardless of their solid masses, similar ones on the Moon would likely be less alien, considering that the density of matter on the surface of our satellite is lower than on Earth, and the lesser forces than our volcanos there must have much greater effects.

So let's hope for some valid reason, one day, to learn something about the events of nature on the Moon, we must never tire of observing, drawing, and especially measuring.

Any hypothesis without evidence by observation is unimportant in this area;

It becomes harmful if it gives a certain direction to recollections, at the beginning of the observation.

The observer as such has no direction whatsoever in his examinations, except for those by virtue of which he deals only with the proportions of the dimensions, at most with those of light and colour.

Can one at any time assume for certain that, according to many years of observations of one, certain flat area on the Moon of for example 100 square miles, not a single clearer crater or hill is present, it will later be justified to say that you have discovered a new formation on the Moon, when at the imaginary spot a mountain of only 20 toisen high is found, because this is not only conveniently visible, but also measurable.

Furthermore, if it is possible for us to indicate the maximum height of the mountain Calippus to be 2980 toisen with the certainty that the probable error is only +/– 20 toisen, then should one later in time discover that the mountain height to be 200 Tosien larger or smaller then with mathematical certainty one may claim that this mountain has undergone a noticeable change.

The measurement alone is what decides.

The maximum height of a lunar mountain above the average surface can be assumed, without exaggeration = 4200 toisen, the largest depression of a crater floor under the average surface of the Moon to be maximally 1600 toisen. The sum of both makes 5800 toisen or 34800 par. Feet, or 1/154 of the Moon's radius.

This number applies to the large height difference, while the (approximately) absolute height of a mountain of 4200 toisen is only 1/212 of the radius.

For the Earth, the corresponding numbers are as follows:

Highest absolute altitude = 4406 toisen or 26436 par. feet (Kangchenjunga); largest sea depth discovered in 1852 = 7230 toisen;

The sum of both = 11636 toisen = 69816 par. feet = 1/281 of the Earth radius, while the absolute height of said mountain is only 1/743 of the Earth radius. [74]

Ed: Until 1852, Kangchenjunga was assumed to be the highest mountain in the world,

Chapter XVI
On the Distribution of the Plains and mountains on the Moon

Although there are neither large nor small accumulations of water on the Moon, at the first sight, there is a very striking difference between the configuration of large surface areas, which encompasses many thousands of square miles, forming layers of dark coloration, and between the lighter and more hilly regions surrounding them.

This difference in itself does not show any resemblance between the Moon and the Earth.

If the bottom of the oceans and inland waters were emptied, then the bottom of the sea would appear to be an uneven land, partly occupied by high mountains; on it one would also see flat surfaces of great extent, which are to be compared with the flatlands of America and Europe, but in consideration of the colour, the contrasts which occur on the Moon are not to be expected.

The colour of the plant-less desert is different, unlike that of the uniformly heather-covered plains, the grassy llanos, or the vast forest landscapes of South America.

On the Moon, all the larger connected flat lands have a dark grey colour, with green and brownish shades;

if one abstracts from the surfaces of walled mountains and from the even, bright interspaces of many mountain ranges, then the grey areas (mare), formerly thought to be seas, are chiefly found in the northern hemisphere.

Seen as a whole, they lack no mountain form, but the bumps are rarer in it, separated by large spaces, and colossal masses occur only at their edges.

Worm-shaped, 10 to 200 toisen high grey mountain veins are widespread throughout them, as well as bumpy terrain, weak trough-shaped depressions, small and medium-sized craters, shiny isolated rising mountains of moderate height, rilles and groups of smaller hills.

They are also traversed by streaks of light, which, without altering the plane conditions, have their beginning at the ramparts of mighty ring mountains.

Where high mountains appear on the edge of such plains, the rugged higher side often faces the plain;

J. F. J. Schmidt, *The Moon*, Historical & Cultural Astronomy,
https://doi.org/10.1007/978-3-030-37269-9_16

Fragments of the grey colour appear scattered as dark spots in individual ring mountains, e.g. in Alphonsus and Petavius, or the grey covers the whole inner surface of a crater, with the exception of the wall mountains.

It is believed that in view of the rarity of the mountains in the Mare, these grey plains still represent, as it were, a picture of the ancient surface of the Moon, which was later shattered by inner forces, by breaking lighter masses for the most part. They were able to resist more, and where they had to give way only a few of them remained in the general devastation here and there. In them, the force acting from below was often compelled, after futile attempts to break through [75], to wedge a horizontal direction, in which it heaved miles and miles of the ground to flat mountain veins.

The weakened force, if it was still able to do so, finally created a small crater at the end of such vein.

These considerations, not without good reasons, can be carried much further, but the observations are not yet sufficient.

Leaving this hypothesis, let us turn to a casual estimate of the area of the grey plains. If the visible hemisphere of the Moon is divided into 4 quadrants, as is the case in the MÄDLER map:

the northwestern quadrant the 1st; as viewed at the top right
the northeastern quadrant the 2nd; as viewed at the top left
the southeastern quadrant the 3rd; as viewed at the bottom left
the southwestern quadrant the 4th; as viewed at the bottom right

It is found that the two northern quadrants 1 and 2 contain most plains.

The biggest difference, however, takes place between the 1st and 4th quadrants.

The surface area of several grey plains has already been calculated by MÄDLER.

I have added a few others to these difficult and very daring estimates to approximate the total contents of all dark surfaces, but only round sums are taken into account.

Mare Crisium	3000 Square Miles
Mare Vaporum and Sinus Medii	4000 "
Mare Serenitatis	6000 "
Mare Tranquillitatis	6000 "
Mare Fecunditatis	7500 "
Mare Nectaris	2500 "
Mare Frigoris	5000 "
Mare Humorum	2500 "
Mare Humboldtianum	2000 "
Mare Imbrium	16000 "
Mare Nubium	18000 "
Oceanus Procellarum	90000 "
The remaining scattered grey areas	5000 "
Sum	167500 Square Miles

These of course, contain all transitional formations, medium-high hills, Light rays, isolated mountains, and craters.

So as not to exaggerate, we set the total area of the plains only = 160000 square miles.

Since now half the surface of the Moon-sphere contains 344600 geographical square miles, it is evident that all the grey plains together do not yet cover half the hemisphere.

MÄDLER's estimate is only 2/5 or about 130000 square miles.

Similar to how on Earth, by the distribution of land and sea, the character of a landscape, or, in a geognostic and geographical sense, the character of a whole part of the world may be determined, it may also be said of the Moon that the general impression afforded by the sight of its surface is conditioned by the shape and distribution of the grey plains visible to the naked eye.

This is true for the full Moon, because at the time of the phases sometimes the grey areas do not stand out significantly, and it is the predominant type of Ring mountains that gives the Moon such an unfamiliar appearance.

When we look at a landscape, our judgment about it depends on the sensuous impression which the formations and colours evoke; we speak of the greatness, the gravity, or the grace of a landscape, according to what is seen and how it touches our mind.

Soon the richness of the plants and animal forms, or the alternation of the cultivated land with the forest-covered mountains, sometimes the dark line in which the sky rests on the horizon of the sea, or the smoke column of a distant, cone-shaped volcano, the reflections of the lakes, or on high Felshorn the lustre of the eternal snow shining brightly over the cloud cover, which vividly captivates our attention, arouses our admiration, and provides the searching eye the great features that characterize the landscape.

If, for our sake, the Moon does not present such a wealth of views, the eye would prefer to adhere to the dark, often jagged surfaces, to the shimmering white rays which interrupt the monotony of those, and on the dazzling garland of mighty peaks, when they cover hundreds of square miles with their parallel narrow and pointed shadow forms with a deep darkness. It always returns to the plains; the charm of their sight, and the quiet, mysterious charms which their development from the night to daylight often presents, is more stimulating and beneficial than the tiring and strenuous observation of the bright mountain-country, where colossal craters are pressed together in ever-repeated forms and without Change of the light and the colour of the viewing nowhere provide resting points. - Drawings and measurements in and around the grey plains are easier and more pleasant than in the wild light-radiant mountains; for the Selenographer, the Mare are an oasis in the desert.

Chapter XVII
Mountains

After highlighting the contrast between the dark plains and the bright mountainous country, let us look at the mountain masses themselves, and again only at their general structure.

The first characteristic feature which stands out in a telescopic view of the Moon, even to the most untrained observer, is the circular shape.

It repeats itself in many thousands of examples, of which the largest mountains may be amongst the largest formations on our neighboring world, but also lesser formations, whose extreme smallness escapes our certain detection.

It is unmistakable on some well-delineated grey areas (Mare) there is the unmistakable tendency towards the circular enclosure, the resemblance to the Ring mountains.

Where several plains touch each other, their often mighty boundary walls in large bays show parts of a circular arc, which, interrupted here, appears to be continued in another mountain range.

Let us agree with this view, which is supported by the nature of the boundaries of Mare Crisium, Mare Serenitatis, Nectaris, Humorum, the large mare-like ramparts Kästner and Mare Humboldtianum - after all, this marvellous inclination to the circular shape, which is distinguished by the greatest differences in the mare, is not to be overlooked, and it is very probable that they should not be regarded as purely accidental.

It gives us an idea that in ancient times the first and most formidable catastrophes on the Moon had shaped these very forms, many thousands of others of which have followed on a smaller scale and in innumerable passages.

It will be advantageous, for the sake of clarity, first to describe the actual crater-mountains, and only later in particular the mass-mountains (Massif) and the isolated heights of the Moon.

J. F. J. Schmidt, *The Moon*, Historical & Cultural Astronomy,
https://doi.org/10.1007/978-3-030-37269-9_17

Chapter XVIII
The Ring mountain Formation

Within the section regarding elevation measurements, one has already mentioned the indentations.

We have learned that circular basins, with their bases, are considerably below the surface of the Moon.

Probably in all cases, this depth is surrounded by a more or less elevated wall; the crater Rim, which frequently rises even up to 800 toisen over the surrounding land. [76]

In the case of the larger structures, the ramparts are sometimes not very connected; it does not only appear very jagged and open, but also shows considerable deviations from the circular shape.

In most cases, the depth surrounds a rampart closed wall, the height of which remains nearly the same all around, whose outward slope is between 1° and 4°, but whose internal drop is often very rugged, with walls inclined 20° to 50° to the depth to lower itself.

There is no doubt that the top rim of the regular ring-shaped mountains is very sharp, sometimes topped by considerable peaks or by small undulating peaks and humps.

On the outside the wall is little or no structure, but internally it shows in numerous abundance a system of simple, double or multiple terraces, which, beginning at the bottom of their inner base, ascend in concentric store against the ramparts, and thereby diminishing its steepness.

Two terraces each are separated for the most part by a very narrow gorge, bordered by rugged walls; only now and then do we see bridge-like connections (not arched passages, but compact transverse dams), or larger elevations, which have affected there the regular valley formation. [77]

The higher the terrace, the narrower and rugged its uppermost edge; below, the terraces become more irregular, rugged; in the form of very small hills they go down into the crater floor, or are even connected with the Central mountains.

Almost consistently, where a disturbance occurs in the circle design and even in the terraces, it has been affected by a crater of medium, but more often of smaller size.

J. F. J. Schmidt, *The Moon*, Historical & Cultural Astronomy,
https://doi.org/10.1007/978-3-030-37269-9_18

This disturbance appears only as an interruption in the mountain range, and in some cases also (although this is not yet decided by observation) as a lateral displacement and displacement of a mass which may formerly have been in the place of the newly-formed interruption. [78]

Where many craters lie side by side, one can see many places very easily, which are of recent origin;

the destruction which small craters have done to older formations are often most obvious, and no less, when they touch or interlock with their ramparts, the changes of the circular shape, which often degenerates into strange shapes, so that the opinion is obvious; the entire mass had not yet completely solidified when the later disturbance occurred.

While it is generally clear that most craters form deep basins, they are hemispherical cavities; one also finds very strange exceptions.

For, in some cases, the most obvious is that the base of the crater is not only not deepened, but is strongly expanded (distended) or dome-shaped.

On such a bulge lies a central mountain, as in Petavius, [79] or a series of very small craters and hills, as in Mersenius, or a feeble system of very low mountain veins, as in Wargentin.

The measurements further inform us that in certain cases the interior of the Ring mountains does not form a crooked or any deepened surface, and indeed even a surface lying above the level of the Moon.

All these differences of form, which must be subjected to a much more special investigation, point to the variability of strength and of local resistance, whose result we see in countless craters clearly in the telescope in front of us

The name of the various ring mountain forms, as applied by SCHRÖTER, LOHRMANN and MÄDLER, is to be maintained in any case; I also follow it in general, without being particularly strict about it.

Consequently, the various types of the Ring mountains can be better considered, if one distinguishes them according to the characteristics to be described now.

1. Old walled plains, to which I at least count the closed Mare-plains; they are more than 30 miles in diameter, are internally comparatively very even, perhaps recessed below the average surface of the Moon, and generally circular; e.g. Mare Crisium, Mare Kästner [*now known as **Mare Smythii***], Mare Humboldtianum.

2. Ordinary walled plains, wholly or more frequently of only approximately circular shape, between 30 and 14[Spelling - probably 40] miles in diameter; the walled mountain is very rugged, from very unequal height up to 1500 and 1900 Toisen.

   The terraces are missing either entirely, or are only very incompletely developed.

   Youger craters and later rilles to be described have often damaged the wall mountains, so that it appears to be a coherent whole only in certain lighting. The surface enclosed by the ring mountain is little or not at all deepened, light or dark, sometimes also colorfully marked [80], filled with very low hills, small craters, and furrowed by rilles.

   Examples:

   Plato is very circular, internally very evenly deep grey and highly flat. One can see in the inner surface 5 to 7 very fine hills or crater, which cannot be exactly determined. The bright mountain range is divided into individual, but still cohesive masses of very unequal height, which nowhere exceed 1400 toisen east and west.

Ptolemaeus, over twenty miles in diameter, with a very rugged wall, which deviates considerably from a circle.

The height difference is between 2000 and 600 toisen.

In the west lies a Massif, which together forms the broad wall of Ptolemaeus and the Albategnius, rising above 2400 toisen.

In the north-west, a mighty mountain, like a cap, strikes the plain, the height of which is found to be 1780 toisen with the Sun very low; later the height is only 1300 toisen, and from this, as from the diffuse form of the shadow, one can see that the summit must be flattened. [81] In the north-east rise the wall mountains to 1500 and 1600 toisen. In the plain, there is no central mountain, but very many small craters, the largest of which is in the NW, is easily visible in any lighting. Its rampart rises up to 210 toisen above the plain. In this I found in a cursory observation using the 14-foot refractor of the Berlin Observatory [Fraunhofer-Refraktor – used to discover Neptune in 1846] using 250 magnification, at least 46 small craters, of which about 12 on the eastern edge, at the inner foot of the wall, strung together like a string of pearls. [82]

A similar but much more easily recognizable array of 6 small craters lies northwest outside of Ptolemaeus.

The sunrise over this strange walled area afforded a splendid sight.

With sunlight at a 1 to 2° angle, it appears consistently rough and uneven; only at a height of 3 and 4° do you think you can see them as smooth as glass, and later on, the little craters stand out clearly.

The 18-mile-wide walled area Alphonsus, which borders south of Ptolemaeus, has a quite different character; it is more circular, the wall more closed, summit rich, and surrounded internally with very imperfect terraces.

In the middle of the area rises the 700-toisen high Central mountain, connected with long low hills, which form partially grooved valleys.

A perfect, very remarkable rille lies in the western part of the plain. [83] All these phenomena recur more completely and sharply in the beautiful ring-shaped mountains of Arzachel, south of Alphonsus, which must already be counted among the large and deep regular craters. - Other substantial walled areas (wall surrounded plains) on the Moon are for example Grimaldi, Riccioli, Schikard etc.

3. Transitional forms: a daring expression, which can be justified in some examples, but not in others.

If I designate a non-recessed, irregularly circular walled plain, such as Ptolemy and Schicard, as a normal one, then I see the transitional form (without, however, wanting to attach particular importance to this distinction) in those mostly colossal structures, the inner surfaces of which are considerably sunken, and as a result of which the crater form inclines, and because of the great irregularities and destructions of their mountain crest, are still very much related to the walled plains.

As such transitional forms (to limit myself to safe examples), I describe Clavius and Maurolycus.

Clavius is over 30 miles in diameter, surrounded by a huge, several miles wide, partly terraced, very high wall, and so deepened that, while a south-western summit is about 2700 toisen higher than the depth, whilst the western slope of the wall is just 750 toisen, and rises only about 950 toisen over Clavius of the western bounding hilly terrain. Many medium and small craters (at least 90) have broken through the plain as well as the mountain wall, and have partly smashed it. [84]

The whole west wall, which is inclined only by 10° due to the irregular terraces, rises above the interior surfaces to a height similar to that of the Montblanc summit. If, for instance, one includes the greater landscape of Bailly [basin], Clavius may be regarded as the largest ring-mountain of the Moon.

Maurolycus is scarcely half as large, though circular in its entirety, but the rampart, partially destroyed by intervening large craters, is girded all around 1900 to 2200 toisen above the depth; to the southwest the wall appears at a truncated height of at least 2800 toisen, which drops to the west only slightly. [85]

In the depths, there is a central ridge about 520 toisen, and along the whole length of the outer west rampart a groove-like valley is observed, which is probably still drawn into Barocius.

The adjoining great ring-mountain Stöfler, and the mighty wall ring, which surrounds Fabricius and Metius, which MÄDLER has left without naming, and finally Longomontanus can be included in this class.

If these mountains showed signs which more readily suggested their so-called selenognostic class, there are others in which this is not the case; they form a separate class, in which the characteristics of the walled plains, of the transitional forms which have just been described, and of the great terraces, are represented.

I mention: Archimedes, circular, terraced, little or not at all deepened, a little bright inside and quite smooth; Capuanus, Pitatus, Fracastorius, all three large, incompletely closed, without clear terraces, appear as dark bays of the Mare, as if only in a part of the circumference the resistance of the Mare Plain would have allowed the breakthrough.

Their surface is dark and not deepened; the ramparts were very jagged and of unequal elevation.

Gassendi, 12 miles in circumference, circular, partly terraced, with central mountains, hills, and numerous grooves in the inner surface, which is not only not deep, but is probably higher than the Mare [86] -Also the already mentioned Alphonsus, maybe Albatengius, Posidonius and many others

4. Big craters with complete and multiple terraces and central mountains. — MÄDLER attributes most of these forms to the walled plains; the features that I will point out have caused me to deviate from the previous designation.

These features are:

Most substantial depression, approximately a circular form;

mostly wall-like rampart, which on the outside lowers and often very irregular, inside with double to five-fold terraces down to a depth in quite a few cases.

In the middle of the basin almost always rises a simple or more frequently multimodal [many peaked] Central mountain.

It is a special feature of these often 12-13 mile wide and 2500 toisen relatively deepened crater that they have been affected little by small later craters. They are also characterized by the great brightness of the upper hemisphere, by the system of rays (which will be described later), and by the distinguished radiating outward chain of hills. All these properties are clearly demonstrated by the following examples:

a) large crater of 9–12 miles in diameter, with numerous terraces and central mountains, with very slight disturbances by later small craters, and with great light radiations: Tycho, Copernicus, Olbers, Anaxagoras; in the latter, the terraced system as well as the Central mountain are less developed.
b) Similar shapes without light rays:
   Moretus, Werner, Geminus, Bullialdus, Eudoxus.
c) Similar forms with a radial hill system:
   Aristillus, Autolycus, Aristoteles, Langrenus, Theophilus

A ring-mountain, like Copernicus, unites all the features of this class; careful studies, only on this incomparably beautiful and splendid structure, are wholly absorbed by a hundred other craters.

It sends its broad rays of light through the grey plains, and miles outwards, hills of medium brightness, which narrow valleys often enclose in a parallel direction.

These begin only at some distance from the outer slope of the rampart, which, not much raised, even on the outside, has concentric terraces and hills of very great numbers.

The highest, especially in the east partly meandering wall is shiny and steep, lies almost two thousand feet above the base, and descends with inclinations from 50° to 60°, downwards with inclinations from 10° to 2° to the bottom. The inclination of 50°, and in some points 60°, is peculiar to the highest ridge, and to some terraced walls; the high peak rising from the western ridge, with its summit above 2100 toisen, is higher than the base of the many small central mountains, which are not yet 400 toisen high, [87] rising above the upper western terrace still 1080 toisen, the arched ridge of which is 937 toisen higher than the base.

The probably concave floor contains many central mountains and hills. Among these surveys there is no particularly striking and high central grouping. The eastern central mountain is 370t, the western 303t high.

A crater is not found in its surroundings; In fact, the entire Ring mountain is missing the later breakthrough of a definite crater. No one has yet succeeded in determining the height of the western wall.

For the eastern wall, which does not reach 2,000 toisen above the surface, I find from 3 poorly harmonious measurements the uneven and inclined ground to be a height of 670t. The outer northeast terrace has a height of 470t.

If Copernicus is just 5° to 10° from the terminator , it is clear from the effect of the contrasting illumination not only the elevation of the main wall(which is already known),

but the slow ascent of a 4 mile wide stretch of land to the wall, a general elevation of the terrain, about 22 miles from the west to the east, from which the ring-mountain range has broken. - Copernicus is very similar to Langrenus, which is much less suitable for a precise investigation, due to its western position. On its Southeastern Wall, elevation differences from the Base are to up to 2400t, and two great central mountains may be more than 900 toisen. There are places on the inner upper western walls, which may be inclined by 50° or more.

Tycho, which lies entirely in the bright mountains of the southern hemisphere, has a completely closed wall, and in the north, is probably a height of at least 950t.

Theophilus, towering over the valley to the west and east by 2200t, shows in the Ring mountains, as in the broad terraces, already small crater breaches, a multi-peaked, more than 1000t high, central mountain, and many interacting hills, which are connected with the eastern terraces.

The outer wall height does not exceed 500 t. [88]

5. Medium, regular craters. They share many properties together with the previous class; their average diameter can be assumed to be 5 to 8 miles. Deviations from the circular shape are irrelevant; they have neither terraces nor central mountains.

The floor is probably concavely shaped in most cases. All are distinguished by considerable brightness, such as Aristarchus, Kepler, and Dionysius, upon whom, moreover, the radiation system is more or less strongly developed. The terraces, however, are generally confined to one, or at most two, concentric steps, and the central mountain range often rises only in one peak.

Examples of such craters are:
Menelaus, Manilius, Kant, Reinhold, Landsberg, Timocharis, Lambert, Pytheas and many more

Smaller in size and subject to extremely low central highlands are given: Bessel, Picard, and Theophilus A.

To which of the latter, in the northernmost part of the Mare Nectaris, I intend to add the name of the famous Dorpater astronomer, Mädler, who is highly knowledgable of the lunar surface.

Among these craters, which are distinguished by the floor, interconnection, and ruggedness of the wall, and which are, in many cases, comparatively recent in origin, strange phenomena appear in rare cases.

In powerful telescopes, one sees in some detail that the basin contains a second crater, which with its rampart not only touches directly the inner foot of the main wall, but also equals its height. The bigger crater encloses a completely similar, but slightly smaller one. MÄDLER first saw something similar in crater Vitello, but here the wall surrounding the Central mountain is slender and low, even lower than the Central mountain, and still far from the main wall.

I found the strange configuration mentioned above in 1853 using the Berlin Refractor on two small craters, one of which destroyed the wall of the Hesiod(Gesiod), the other lying west of Ramsden in the plain[89] , under the most favorable circumstances, can also be seen using a five-foot refractor with a magnification of 200 to 300 times. - In the case of the medium-sized craters, walls inclined inward from 30° to 40° are not the rarities.

6. Small crater. Their number is indefinably large, and the Moon's hemisphere on this side shows probably at least 50,000 of them. [90] Their smallness does not permit us to look at their construction or determine their depth.

They occur without exception in all regions; they have broken through everything that left the old cataclysms intact. They lie in the plains often near each other, and in long lines, as if broken as a column; so between Eratostenhes and Copernicus, where among the approximately 300 craters visible there, many touch their ramparts, and so often intersected in such a way as to form groove-like narrow gulches.

Their outer height is very insignificant, between about 20 and 90 toisen. [91] These small craters are sometimes found on the highest crest of large ring mountains, but very seldom on the summit of high mountains, and excepting one or two cases, crater-mountains, like our Etna or Cotopaxi, are by no means to be found on the Moon. [92]

Whether these smallest craters (the after effects of previous greater forces) still arise today, cannot be satisfactorily demonstrated.

No single field of the Moon is sufficiently investigated even with a five-foot refractor; and how difficult it is to recognize many of the smallest craters is well known to the practiced observer.

Anyone who can appreciate all the circumstances which may be considered in the observation, who has a complete overview of the works of SCHRÖTER, LOHRMANN, and MÄDLER, and has for many years made the Moon the subject of very special investigations, must admit that perhaps not a single case, which gives a reliable account of a new formation. [93]

It is not at all improbable that a small crater might appear; to assume that nature has completely come to rest on the Moon is quite groundless, and all the more so since we cannot prove a strict stability of the minima of visibility.

It has to be our task, after two excellent General maps have already been provided by LOHRMANN and MÄDLER individual and less extensive, not markedly affected by libration for many years under all circumstances, with the most powerful telescopes, and what is so very desirable to observe under a favorable sky, to measure and record what we have seen to great accuracy in specialised maps.

The future will then decide if and what changes have occurred. But to look only for changes before a sufficiently detailed topographical map is available can only be undertaken at this time by those who are less able to cope with the subject because of a lack of insight into all the circumstances to be considered.

7. Rille. If these marvelous forms are characterized as long and narrow furrows, as broad, wide depressions, or as cracks in the ground, for instance, by the cooling

of a temporarily strongly heated surface, it may be strange to associate these structures to the craters.

Alone, the Transitions are here too, which indicate to us how the grooves are to be understood

MÄDLER was the first shed light on this subject. [94]

He showed that single rille not only pierce through the walls of some small craters, and with or without their own edges, but that a rille also appears as the result of many juxtaposed craters whose walls are broken in one and the same direction, a long furrow surrounded by numerous semicircular crater walls on either side. - It may well be thought that "the grooves are the result of the forces which the crater has created."

The fact that individual grooves do not show the phenomenon mentioned, that they appear only as very narrow, slightly curved crevices, is fundamentally not a strong counter-proof, as it is conceivable in many cases that the eroding force in many points was so close to each other that so quickly that no time was left for regular cratering.

But next to that, really straight crevices may occur, or the collapse of an impaled wrinkled ridge; all these assumptions can be confirmed more or less by examples.

So much seems to be certain that, according to the nature of the appearance of the grooves, especially in the mountains, as they continue to pass through craters, regardless of their grade, and perhaps and maybe in some cases split whole mountains, they appear as the most recent formations the Moons, which probably have not yet completed.

If, according to the present state of our knowledge, certain more or less doubtful forms are included among the rilles, the discoveries of the same are distributed as follows; they are gradually found:

5 rille from SCHRÖTER	since 1788 [95]
95 rille from LOHRMANN	since 1820 [96]
44 rille from MÄDLER	since 1830
3 rille from GRUITHUISEN	since 1822
5 rille from KINAU	since 1849 [97]
97 rille from SCHMIDT	since 1842.

The largest share of the rilles first seen by LOHRMANN was, however, independently discovered by MÄDLER, as I have only recognized after the insight of the still unedited maps of LOHRMANN. At the time of MÄDLER's observations, only four sheets of the Dresden Selenographer had been published.

The 97 rille found and drawn by me remained unknown to the former observers. These figures are, however, not to be taken with all rigor; some forms are so dubious that the decision can only be made by means of larger instruments. In any case, however, it is certain that the larger telescopes show at least 200 grooves on the Moon. In order to see the least distinct rille, it is necessary to increase the magnification, have a completely calm air, and, what is more important, an eye familiar with the observation of the Moon. - Some of the rille I know have been seen only once.

From the regions where the rille are numerous, I note the following, and also some particularly remarkable ones:

1. The Mare Vaporum and Mare Tranquilitatis, where, including a few mountains, the rille that begin west of the Sabine in the Mare Tranquilitatis, through the extraordinary areas of the Ariadaeus and Higinus [Hyginus] to Triesnecker, already noted by SCHRÖTER and the northern districts of Ptolemaeus, without verifiably hanging together everywhere [98]
2. The area Arzachel and Alfphonsus, where it occurs within these large ring mountains in sharply drawn but difficult to recognize figures. Here you can also see them in longitudinal valleys. Inside the large walled areas you will find further rilles:

   in Vendelin [Vendelinus], Petavius, Posidonius, Grimald, Pitatus and especially Gassendi, in whose area alone MÄDLER discovered 14 rille, of which 7 can be seen using 5-foot refractors.

   Probably, in total, there are 18. [99]
3. Perhaps the strangest of all rille systems is in the grey plains surrounding the crater of Ramsden; it is connected with the Hainzel by debateable forms, and probably with the long rille, passing through the Mare Nubium to the Pitatus; perhaps also with the distinct rille system of the Hipalus. Ramsden is the center of four main furrows, which are joined by seven other curved crevices. I first saw them in Bonn in 1849.
4. Transitions/Passages between crater and rille are found between Eratosthenes and Copernicus, at Piccolomini, perhaps in the Ptolemaus, and, for example, in the large Higinus rille.
5. Very curved, hard-to-see rille, previously unknown, lie at the foot of the Apennines at Hadley. [100]
6. The rille system around Mersenius, where various highly remarkable and easily recognizable rilles have remained unknown to all earlier observers. Here, parallel grooves follow straight directions, set out on the ramparts of large crater, without breaking through, when they have only attacked the mountains (and this is mysteriously indicated by a grey band-like shading), or have modified (with regard to the ability to reflect light), they penetrate the entire bottom of the crater in the same direction, and in one case such a rille is cut at right angles from another.
7. Numerous rille at Grimald, Lohrmann and Hevel [Hevelius].
8. The great Sirsalis rille system, 60 miles long.
9. The curved rille at Aristardi [Aristarchus]
10. Rille south of Newton and Cysatus. [101]

Only some of the most important localities have been mentioned here. — One will probably soon be convinced that the rille are nowhere to be found on the Moon, and perhaps not many years will pass that judicious observers will decide whether or not rille are still being created.

Much can be explained by the variability of illumination, of libration, and by many circumstances which influence an observation; but not all.

In order to appreciate the dimensions of the rille, it is noted that they are broad at 4-20 miles long, 300 to 2000 toisen wide, and 50 to 200 toisen deep. [102]

The beginning and the end are mostly indicated by nothing, sometimes by a crater; the breadth is not very variable, the inclination of the inner walls is not very steep, and the mutual margin seldom distinguished by an elevated recognizable wall.

One will not be inclined to consider the rille as Selenite structures.

If one does this, one implies implicitly purposes which are well-founded in buildings on Earth;

We subordinate these purposes to creatures whose qualities and needs we not only know nothing about, but whose existence is also completely unknown to us.

According to our conceptions of the purpose and the usefulness of earthly man-made works, we should regard as very foolish a number of rille, as far as they are held to be a building.

Chapter XIX
Mountain Ranges and Cordillera (Chain mountains)

The prevalence of the Ring mountains on the Moon is so significant that the areas where they appear rarer (except in the plains), and where very large masses of land are piled up without special order, comprise of only a few cases of a larger areas. Mountains of the form, such as our Cordilleras, Alps and Himalaya, are absent on the Moon; the resemblances are not significant, taking into account the group as a whole, the predominant directions, and the situation of the highest peaks.

The South American Cordillera form a double and multiple chain, which slope off unevenly on both sides, containing between them flat valleys of very great height and considerable extent, and partly on their ridges, partly set side by side, carrying single peaks of over 3000 Toisen.

In the Alps one recognizes individual central masses, of which multiple ridges separated by glaciers, and in the Himalaya, starting at the over 4000 Toisen high snow mountains, one distinguishes, descending to the south, certain regions with respect to height and geognostic makeup, before one reaches the bottom of the Indian plain, in the area of the sandstone and the marshland (Terai - a lowland region in southern Nepal and northern India).

The character of an earthly mountain can be understood in very different ways. We have to examine: the ratio of the mean crest height to the summit height, the mean direction of the main ridge, the Dip direction and the Strike of the strata, insofar as the mountains are not exclusively plutonic rock; the mineralogical makeup are determined, by assessing, that for example the snow peaks of the Cordilleras from Trachyten and Andesiten, [103] the Alpine peaks, consist of Gneiss and limestone etc. [104]

One separates the more or less random grouping and series position of a mountain from the characteristics of the rock-types themselves, which indicate very different epochs of the origin or the transformation.

J. F. J. Schmidt, *The Moon*, Historical & Cultural Astronomy,
https://doi.org/10.1007/978-3-030-37269-9_19

We cannot make all these distinctions in the mass mountains of the Moon; It is left to us alone to consider the external form, the extent and direction of individual mountain limbs, which can be recognized on the lunar surface, and the altitude conditions.

The Apennine mountains of the Moon cover an area of about 3800 square miles. [105] It stretches from the great crater of Eratostenes, in a north-westerly direction, on the edge of the Mare Imbrium, in order to unite with the valleys of the south-through numerous limbs it branches south to the plain of the Mare Vaporum.

In general, it cannot be denied that in the whole mass of the rock there is a coherent, approximately from south westerly – north easterly direction of many ridges. [106]

From the Mare Vaporum, the mountains rise quite gradually, and end in the splendid very steep valleys at the Mare Imbrium, where the highest peaks rise. Their heights above the Mare Imbrium are often greater than 2000t. The long ridge of Huygens rises 2800t across the plain east, but only 850t over the Apennine highlands in the west; Bradley 2100t over the Mare, 1730t over the western neighbouring mountains.

But even the northern edge of the Apennines is not connected; separated by frequent gulches, it also disintegrates into single masses of considerable extent, and it is now impossible to give numbers which certainly express the ratio of the mean crest height to the summit, and which in any case would only be valid for the northern precipice.

This slope is now surrounded to the north by low hills, which sometimes rises in steps at the foot of the main wall, analogous to a mighty mountain of rubble, which fell back upon the rising of the masses, and remained in its position at the foot.

No matter how many valleys and short, often curved, canyons are seen, nowhere do we see the long-necked valleys so characteristic of the mountains of the Earth, which conditioned the course of the rivers, forming either simultaneously with the formation of the mountain or in the course of time were cut through so deeply and steeply by the water.

In the Apennine of the Moon there is indeed a highest margin in the north, but it does not occur anywhere except with the sharp wedging of the upper ridge in a certain wall, dammed on either side. In the Apennines of the Moon there is indeed a highest border in the north, but nowhere does it occur with the sharp ridge of the upper ridge in a certain wall sloping roof-like on both sides.

It contains, indeed, the highest peaks, but these, with their immense bases, produce masses in themselves, as the peculiarity of their context and their local separation, are hardly found on the Earth. It is very similar with the Alps on the Moon; [107] they rise gradually from the west towards the rugged slope of the Mare Imbrium, and a very peculiar crossed-crevice, a rectilinear rille of huge dimensions, thus one can prove no long valley at all, but only the random grouping of large and small mountain masses

The average peak height of the eastern edge of the Alps can be assumed to be 1400t; but from a 300-400t high average ridge height can only be used, considering the saddle-shaped depression between any two mountains, for there is no more general middle ridge here than in the Apennines, much less in the neighbouring Caucasus, east of the Mare Serentatis. [108]

Even in these great of massive mountains, so we only see a lot of isolated masses close to each other; there is no continuous development of the longitudinal development, and the distinction between the central elevation of the mountain and the culminating peaks has only a definite sense if it is limited to every single group In this sense one can assume the ratio of the ridges to the summit height: (according to very casual estimates, because measurements are lacking)

in the Northern edge of the Apennines	...	=1 : 1.5 to 2
in the Eastern edge of the Alps	...	=1 : 2 to 3
in the Eastern edge of the Northern Caucasus	...	=1 : 2 to 3 [109]

With regard to our greatest mountains on Earth, on the whole, the occurrence of the highest peaks is restricted to certain regions, when isolated volcanoes, for example, Mount Etna; they either rise from the highest ridge of the mountain, or stand close to side spurs, which run out at right or small angles from the main ridge.

For example, in the Cordillera of Quito, in the western mainland, the peaks: Cotocachi, Pichincha (2490t), Atacazo, Corazon (2470t), Illinizas (2727t), Carguairazo (2450t), Chimborazo (3350t) the volcanoes and other high mountains such as Imbaburu, Ilalo, Ruminavi, Tunguragua, Antisana, and Sanguay either to the west or to the east of a ridge, or to rise on burrs which, in the direction of morning to evening, connect the eastern chain of the Cordilleras with the western.

Consider only that part of the Himalayas which, according to the map of HODGSON [110], extends from the 78th to the 94th degree east of Greenwich, as from the west-east trending well-connected main trunk, 7 marked branches break away, typically, at right angles, and run to the south, where they terminate, before they reach the region of the Bhaver or the Saul Forest.

The mighty snowy mountains contained within the whole mass, such as:

Jamnoutri (4014t), Nanda- Devi (4003t), Davalagiri (4316t.), Gosainlham (3862t), Kintchinjinga (4406t), but by no means are they on the main ridge,which carries only the more eastern summits of Chamalari (3742t) and the twins (3378t and 3358t); but on the above, southward descending transverse spurs [111]

There are no such conditions on the Moon; apparently the high peaks of the Apennines lie on the north bank; but we do not know the heights of the rest of the world a little more southerly; seemingly further, all the large major mountains of the Moon-Alps lie only along the eastern edge, but it is not the case in reality; just as high, and perhaps higher, lie scattered in the West, quite randomly. We might find a great resemblance between the mountains of the Moon and the Earth, if those of the latter had not been exposed to constant exposure to air and water for thousands of years.

On the Moon, we may still see the prototypes of the mountains; not on Earth.

In addition to the three mountain Ranges of the Moon, the Apennines, the Alps, and Caucasus, there are others, but their position does not permit any further observation. We note the following, though less high, mountains, as coherent elevations:

1. Apennine, as a southern edge of the Mare Imbrium, up to 2800t high;
2. Caucasus, east coast of Mare Serenitatis, up to nearly 3000t high;
3. Alps, N.W.-edge of the Mare Imbrium with peaks up to 2000t high;
4. Semicircle of the sinus Iridum, up to 2300t;
5. Carpathians [Carpartus], north of Copernicus, up to 1500t;
6. Haemus, southern edge of Mare Serenitatis, until 2000t;
7. Taurus, western edge of Mare Serenitatis, up to 1300t;
8. Enclosure of the Mare Crisium, up to 2400t;
9. Pyrenees, western edge of Mare Nectaris, up to 1000t;
10. Altai, north-east going from Piccolomini, to 2200t;
11. Riphaeus [Riphaeus], in the Mare Nubium, 600t;
12. The Hercynian mountains, in the Oceanus Procellarum, 900t;
13. Eastern border mountains: Cordillera Rook and D'Alembert, 1000t – 3500t.
14. Southern edge mountains: Doerfel and Leibniz, between 2000t and 4200t.

Chapter XX
Isolated mountains

Apart from the Central mountains already mentioned, which are at the bottom of the craters, in many places on the plains of the Moon, individual mostly clear mountains arise; they are rarer in small groups.

Neither its height nor its breadth and steepness is considerable, and among the mountains of this kind, there are probably none that reaches 1400t above its surroundings.

In the Mare Crisium, in the west, many small hills lie close together, but quite separated, without touching each other at the base; one of them reaches 900t, while on the east side of the surface a bright and broad mountain rises up 1100t. The Mare Serenitatis, Tranquillitatis and Nectaris, as well as the Mare Foccunditatis, lack the more striking isolated mountains of this type.

They are much more frequent in the Mare Imbrium, where the brilliant, but not quite mile wide mountains: Pico, Pico A, Pico E, Pico B can rise up to 1200t, 1200t, 1300t and 1000t height.

The mountain Lahire is 900t, its western neighbour 590t high. All rise very brightly from the dark grey of the plains, without any noticeable connection with other objects.

From the sight of their long and pointed shadows, it is easy to erroneously conclude they are imposing figures, but a profile drawing in which I assumed the width of the bases of Pico and Pico A to be only 11000 Toisen, showed little markings, respectively a 3- and 2-summit ridge, whose shape is reminiscent of any of the steep Alpine heights, and not even comparable to the 1200 Toisen high pyramid of the Niesen on Lake Thun. [112]

J. F. J. Schmidt, *The Moon*, Historical & Cultural Astronomy,
https://doi.org/10.1007/978-3-030-37269-9_20

Chapter XXI
Vein mountains / Wrinkle Ridges

Although faint hills are seen in various bright landscapes of the Moon, the very long, low, and usually summit-free mountains, which are really very long, seem to be peculiar to the grey plains, the Mares.

Their number is exceedingly large, and with the help of a very powerful telescope, one arrives at the conviction that not only are they not absent in any Mare, but that they, except for very small areas, completely cover these surfaces. Few are directly measurable.

One already recognizes those which betray themselves at a height of 10t and 20t by an outline of a shadow; the mountains from 30t to 60t height are measurable under very favourable circumstances, but even the measurement of the 100t to 200t high features rarely succeeds.

Very often they are strongly curved, 10 to 70 miles long, they branch out, ending in craters, or disappear imperceptibly in the plains; others reappear as foothills, as faint folds on the brink of bumpy 20 to 100-toisen high elevations of the floor of the grey plains, or they have a crater as a common centre.

They are all, with few exceptions, just as dark or less bright than the Mare.

In these plains, which so often have a greenish colour over large areas, they are without appreciable consistency, distributed in the prevailing directions; they are as frequently in the grey as in the green terrain.

Sometimes the height is only one-sided; one only notices a long-stepped step in the Mare; as such, the strangely straight and steep wall can be seen east of Thebit in Mare Nubium. [113]

In Mare Humorum you can see between close parallel mountain vein longitudinal valleys, which have much resemblance to rille.

If one succeeds in closely observing certain mountain veins in their development from the moonlit night, one notices their piecewise emergence; one sees that its back is not smooth, but is covered by many small elevations, which can amount to very few Toisen.

J. F. J. Schmidt, *The Moon*, Historical & Cultural Astronomy,
https://doi.org/10.1007/978-3-030-37269-9_21

What makes them visible is neither their height nor their possible shadow, but their horizontal extent.

It may be very easy to see an object on the Moon, which has only half the height of one of our larger towers, if only its width is considerable, and is at least 200 Toisen.

With ordinary telescopes 5 feet in length we would not see a tower of 7 to 10 Toisen thick on the Moon, not even at 100 Toisen height, because the width of the shadow only at angles of resp. 0"007 and 0"010 would be seen.

The greenish colour in the Maren was first brought to attention by MÄDLER, and after his very careful observations he described the boundaries of the green and grey colouring. [114]

In Mare Serenitatis, where the green is best seen, it is quite distinctly distinguished by a generally grey seam, several miles wide, from the bright mountainous land surrounding the Mare. In other places, one notices the green a bit more, but a brownish-green colouring, and in full-Moons, with very good air, and great magnification, the surface of the Mare Humorum sometimes gives the impression that the small bright spots lying in it (crater and mountains) as if through a translucent brown glass, or through such a coloured horn plate. Whether the green area is always the same, whether the intensity and boundary of this colour changes with the lunar phase, or during the covering by the Earth's shadow, cannot yet be decided.

In certain places, the grey is very dark, almost bluish, and you will find, as already mentioned, grey spots in various ring mountains, and in a small, bright crater north of the Langrenus, in which there is a dark triangle, neither seen by SCHRÖTER and LOHRMANN nor by MÄDLER.

With its base, this grey well-bounded area adheres inwardly to the east bank of the craters, running through the considerable depths, and rises with its tip up high at the inner west wall. They are remarkably noticeable at full Moon, and the western tip of the triangle is still visible, when the depth of the craters is covered more than 3/4 by night with a waning Moon. [115] Often Maria, such as the Mare Nectaris, display colourful areas of darker gray, while the grey of some walled areas and craters, such as in Schickard and Buch/[*BYRGIUS]*, in sharp outlines display various forms.

Again, in other cases, Tycho is seen as a grey nimbus: it appears in the bright and even the dark terrain, for instance, in the Byrgius and Aristarchus craters, within the immediate vicinity (with the exception of the effect of contrast) is darker in Aristarchus than the Oceanus Procellarum.

For Conjectural astronomers, who find a great similarity between the Moon and the Earth, since they do not doubt a lunar atmosphere anyway, the existence of the green colour of some plains must be very welcome; they will immediately conclude that the formerly oceanic basins of the Moon are now dried up and covered with rich vegetation.

According to their conclusion, the Moon must have an atmosphere because the Earth has one, and since the weight of certain facts is self-evident enough for their view, they decide to accept the lunar atmosphere as little as possible, so that the objection recedes because of the refraction of the rays.

Because our plants are green, the plants on the Moon must also be green; that is to say, analogical conclusions. They are, on occasion, so sure and self-sufficient in their conclusions that they do not even include certain botanical exceptions in their so-called proofs, which seem to favour their theories; they forget or do not know that on the Earth not only cryptogamic but also phanerogamic plants have been found on the summit of the Montblanc and on the heights of the Cordillera and the Himalaya, which thrive in a reduced air pressure of one-third or more.

We must not blame analogical conclusions because they lack mathematical rigor and power; but they are to proceed from the surely founded, and must necessarily follow the facts only. If we are not satisfied with this, we must at least conceive that there is a boundary between serious scientific research and idle speculation, the latter, seeking the most convenient paths possible and freed from the discomforts of independent observations, and, especially in youthful minds, often hindered or suppressed a strictly scientific, and for this very reason successful, direction for life.

Chapter XXII
Ray Systems

Even in ordinary pocket telescopes, at the time of a full Moon, bright straight stripes of light are noticed, which, from certain points of the Moon, radiate radically out to all sides.

These phenomena, which are the most striking at first sight of the grey plains of the full Moon, and deserve special attention because of their considerable distribution and remarkable properties. LOHRMANN and particularly MÄDLER first examined these stripes more closely. The same condition, unless coincidentally coincident with the position and direction of a ridge, or a mountain vein, does not change the plains anywhere on the Moon.

They appear in the mountains, in the crater depths, and in the grey plains only as variations of the soil colour; they disappear near the terminator, and in their place no trace of a shadow is seen; consequently they are neither elevations nor depressions.

They begin on the outside of large or medium-sized craters, and, regardless of the greatest differences in level, they move in all directions in more or less straight directions, often branching out and sometimes straddling them. They are also recognizable in the brightest mountainous terrain, and it is they who, in the full Moon, through their countless ramifications in the southern hemisphere, obscure most of the great crater forms. In the Mare they are better visible because of the contrast.

We do not have a completely satisfactory explanation, but MÄDLER's opinion will always deserve special attention. According to him, the forces (gases) penetrating from the inside against the surface of the Moon did not always act in a vertical direction; they flowed more under the (hardened) crust horizontally from a large eruption site, to a hole, where they broke through, after they had attacked the surface in manner throughout their course, that, presumably, the structure of the masses changes, but probably their ability, to reflect the sunlight, was increased.

This bold hypothesis (but the only one which is so far compatible with observed facts on the Moon) thus speaks of a metamorphosis in the main, in the sense of the Plutonists; it is confirmed or strengthened by the so-called shaded crater, even by

J. F. J. Schmidt, *The Moon*, Historical & Cultural Astronomy,
https://doi.org/10.1007/978-3-030-37269-9_22

isolated mountains in the grey plains, which show round a fairly broad white glimmer, which either runs gradually outwards, or is distinguished by colourful shapes.

It is sometimes thought that the bright masses of which the crater or mountain consists are seen to glide sideways downwards through the dark grey of the Mare; it seems as if the closer the crater wall was, the more the grey coating has been destroyed.

Nothing can be said about the relative age of a crater surrounded by light radiations.

If the stripes were already present at first, their disappearance does not necessarily follow from the fact that thousands of ring-shaped mountains later broke through the surface of the Moon.

Unless this was completely destroyed, the strips of light remained piece by piece in their old position, and many also see that their course is interrupted for miles.

But if the current configuration of the lunar surface was already completed, one must assume that the light radii of more than a hundred miles in length and ½ to 4 miles in width could still be formed with ease, regardless of the often-colossal height differences of the ground. [116]

While certain large ring-shaped mountains form the centres of such light rays, others are seen, which on all sides emit fine hillocks on all sides; these hills begin at the outside of the valley, are of low height, mostly dark, but arranged in rows, so that each two rows form a long, but not very closed longitudinal valley on both sides.

Rarely do these valleys have the characteristics of the rille. These hill radiations seem coincident with the rays of light. An experienced observer will never fall into the trap of confusing these very different phenomena with each other. You can see how both occur in the following examples.

1. Tycho is the centre of the most important ray system, which we know on the Moon. MÄDLER has described it very carefully, and has supplied a special map, shown in metallurgical (metal screen) prints, in which the development of the bright rays from a grey broad nimbus surrounding Tycho is very well illustrated.

   You can see the Rays in all four quadrants.

   Southeast, south and southwest, they spread in the bright mountainous terrain; north-east, they meander through the Mare Nubium, and to the northwest, though the connection cannot be established, a strip reaching Menelaus, passes through the Mare Serentatis, and turns to the north-western Polar Regions.

   The difficulties of figuring Tycho's entire ray system seem insurmountable.

   All around Tycho is surrounded by large mountain ranges and other wild mountain masses.

   If it lies on the terminator, one sees neither the grey nimbus nor the beginning of its rays.

2. Copernicus shows both phenomena:

   a great ray system, and a richly developed one, especially in the N.W. clear system of radial hillocks enclosing parallel furrow-like valleys; they start a few miles outside at the rampart, and all lie on the already described ground, gently rising around the main wall, the centre of which is called the crater.

We must not, however, suppose that these valleys have a peculiar similarity to the barancos of large excavation crests of the Earth, in the form of long ravines, as in Palma and San Miguel from the edge of caldera (crater), or as in the Pic of Tenerife, and of the Somma of Vesuvius, from the edge of the old craters, which are now no longer entirely present, against the sea, or at Vesuvius, northward against the plain. [117]

3. Kepler, a little excellent crater, the centre of an extraordinary ray system in the Oceanus Procellarum.

   A grey nimbus is absent, and although the hills are frequent enough in their surroundings, they show no trace of a radial position.

   Kepler is surrounded by a broad expanse of bright terrain, from which, on the boundary of the Mare, the long broad streaks of light are drawn, especially towards the east.

   They partly combine with the stripes of the Aristarchus and the Copernicus.

4. Aristarchus, the brightest crater of the Moon, has, though lying in the grey Mare, a darker nimbus (often bluish in the full Moon) around him, from which the fine light radiations extend, and between them some mountain veins. [118]

5. Aristillus, a big, steep crater has a very rich system of the finest radial hill chains, which cannot to be counted individually, but also weak rays of light extending against the Mare Imbrium and the Caucasus, which are clearly visible when the Moon is waning.

6. Autolycus, quite similar to the former, but smaller; shows both phenomena to a lesser extent.

7. Proclus, a very bright crater, is the centre of great light-rays, which spread westward in the Mare Crisium. [119]

8. Anaxagoras, distinguished by the brilliancy and sharpness of its form, is the centre of a large, but the northern position, because of the difficulty of recognizing light radiation.

9. Furnerius *C* and Stevinus *A* (in the fourth quadrant) are still in the area of the large rays of the Tycho, but they have independent, though light, but less clearly developed light streaks.

10. Olbers, on the eastern edge of the Moon, the centre of a probably very extensive Ray system, but unfavourable for observation.

11. Byrgius *A*, a brilliant crater on the west bank of a larger ring-mountain, the centre of a very important, sharply defined radiance of light in the very bright mountains. At the northern latitude of the Moon, and with a strong libration, it appears very sizeable, and is entirely like a spider-web, the threads of which, after heavy rains, are partly destroyed, partly slack, and frequently curved.

12. Zuchius (in the third quadrant), can only be recognized as the centre of the ray when the Moon has a north latitude, and the features are moved westward by the

libration. The crater is very colourful, that is, surrounded by a grey nimbus; its rays meet those of Tycho and Byrgius.

In addition to the twelve ring-mountains mentioned, there are a few others, in which, under favourable circumstances, more or less distinctly developed tracts of the ray- or hill-system are perceived; these are:

13. Dionysius, a very bright crater; it has surrounding it, a strange colourful nimbus, from which some clear light streaks develop.

14. Timocharis in Mare Imbrium, surrounded by weak light streaks and very fine radial hills.

15. Menelaus,

16. Manilius, both wholly and partly in the grey terrain, surrounded by bright corona, which show short radiating streaks. Hill radiations are absent.

17. Langrenus, Petavius, Aristosteles, and Theophilus:

    All four craters of the grandest dimensions are the centres of important hill systems, in view of the localities, in which the radial direction of individual chains is predominant. Only Langrenus shows a few weak streaks of light.

If one attempts to comprehend these remarkable phenomena from a common point of view, we shall always utilise MÄDLER's hypothesis, according to which the forces of the depth, which are directed at certain points of excision, may be regarded as heated gas streams, may one imagine these as heated gas streams or otherwise, the surface of the Moon in terms of the ability to reflect light changed or, in the immediate vicinity of the breakthrough, split the ground when it is lifted, so that the furrows formed all run out from the crater rim.

Where this force was too weak, it confined itself to change the colouring in the immediate vicinity of the site of the eruption, especially in the plains, where some craters and mountains, as mentioned earlier, all surrounding grey colour disappeared.

MÄDLER very vociferously calls this ring, which is surrounded by a white nimbus, "umglänzte crater". ("Aura crater")

It is not yet time to further develop the ideas of these miraculous occurrences, and the abandonment of such speculations is therefore advisable, so as not to enter the field of harmful hypotheses unnoticed with the first indications.

But one may believe that the time will come when the important doctrine of metamorphosis (of the transformation of the rock masses by metamorphic heat) will also be tested on the phenomena of the Moon.

Chapter XXIII
Comparison of Earthly Volcanoes with the Ring mountains of the Moon

It has been repeatedly pointed out that we must take into consideration the external formations in the comparison of the surface configuration of our Earth and the Moon.

The geological distinction between excavation craters and common volcanoes in the sense of L. v. BUCHS [*Christian Leopold von Buch*], as important as it is in itself, we do not need to consider the Moon until a much more observational material enables us to make certain conclusions as to the inclination angles of the Moon's mountains.

Of much less interest to us is the purely petrographic ratio of a volcanic mountain.

We consider only the round, kettle-shaped orifices (crater) of the Earth, and compare them with apparently similar ones on the Moon, not to show internal similarities but more or less characteristic differences.

The superficial comparison of certain circular mountains on the Moon is certainly not suitable for establishing a theory of the genesis of very different forms on different celestial bodies.

At the time of GALILEO and HEVELIUS it was permissible to compare countries like Bohemia, because of their mountain walls, with certain large walled areas of the Moon.

At present our knowledge of the mountains of our satellite is at least so far advanced that we have reason to give up such, incidentally, insufficient comparisons forever.

Let us first consider the craters of the Earth, that is, old elevation craters, ordinary conical ones. active or extinct volcanoes, Maare der Eifel [Maar-diatreme volcanoes], Fumaroles and so on, we will find, in terms of dimensions, very large differences.

We restrict ourselves to the following examples, whereby it should be noted that because of space, and also in view of the clarity, the correct proportions of the height to the horizontal extension could not be faithfully represented in space, they had to be somewhat exaggerated.

J. F. J. Schmidt, *The Moon*, Historical & Cultural Astronomy,
https://doi.org/10.1007/978-3-030-37269-9_23

Fig. 1, 2, 3, 4, 5, 6, 7. These seven examples may suffice to give an idea of the structure of the Earth's volcanoes. From the circumstance that the correct proportion of the height to the width of the base could not be represented, the very steep walls, the great inclination angles of the mountain surfaces, which are much smaller

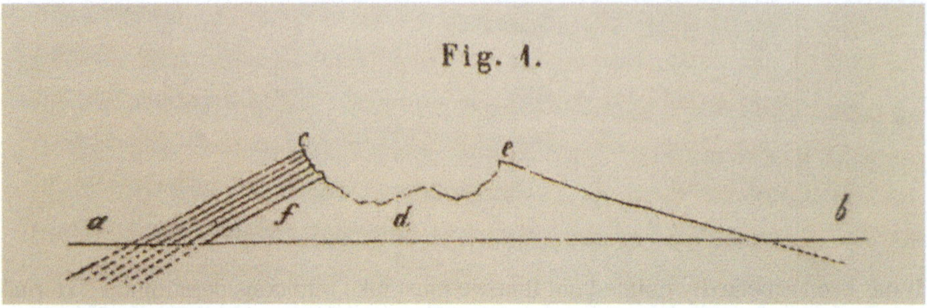

Fig. 1. Cross-section of the Canary Island of La Palma: *a b* sea level; *c d e* cross-section of the crater, the caldera; *c e* crater edges. The oblique parallel lines at *f* indicate that the formerly horizontally arranged mountain strata were raised by the elevation of the mountain. [120]

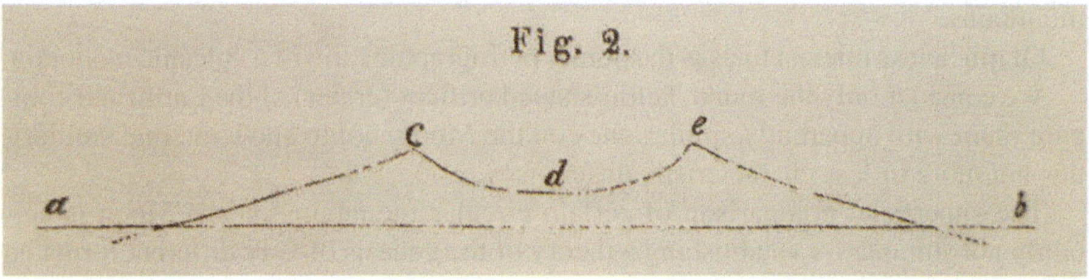

Fig. 2. The suspected cross section of Vesuvius near Naples before the year 79AD *a b* Sea level; *c d e* the old crater, at the place of *c* the current highest summit of Mount Somma, Punta del Nasone.

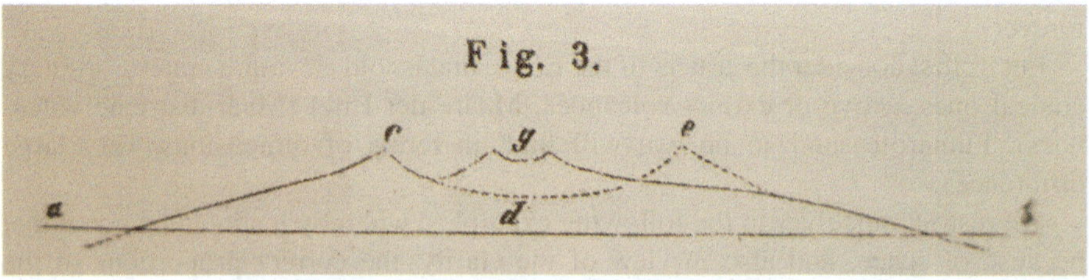

Fig. 3. Cross section of Vesuvius from north to south; valid for the period after the eruption of 79AD; *c d e* the old crater; *c* edge of Mount Somma; the dotted point at *e* is the now-missing southern edge of the old craters *c d e*; *d g* the present Vesuvius cone, as the actual volcano in the old elevation crater; *g* its own crater.

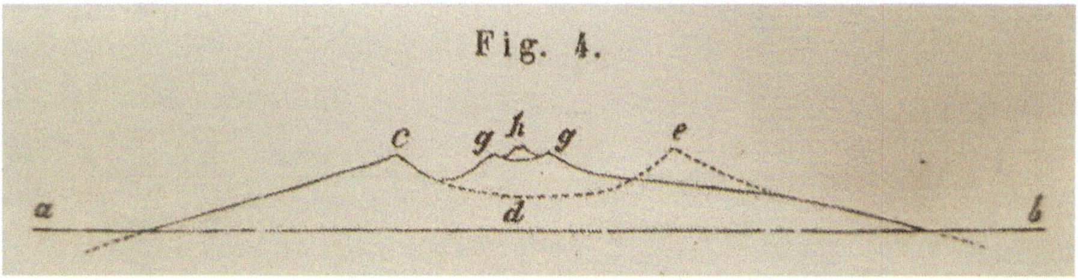

Fig. 4. The same cross-section of Vesuvius after the formation of an eruption cone, which has erupted in the crater of the actual Vesuvius cone *g g*, and, in the form of *h*, extends past the crater edge *g g*. So you have to distinguish: the old structure of the lifting cone *a c d e b*, the later active volcanic cone *d g g*, and the temporarily forming eruption cone h

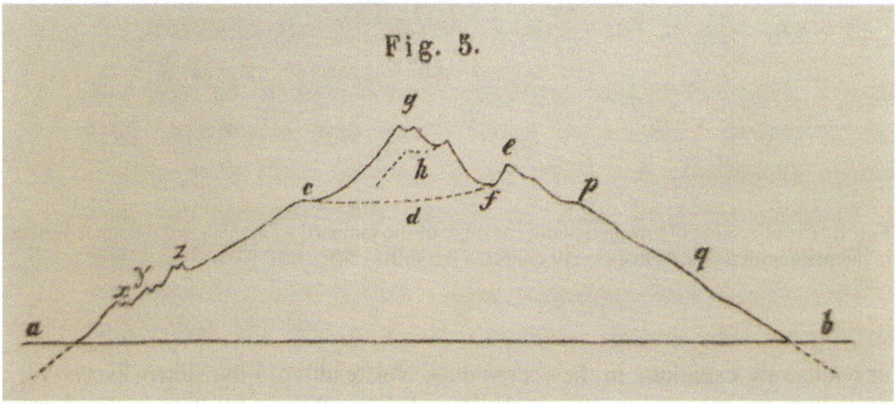

Fig. 5. Cross-section of the western part of the island of Tenerife, from north to south. from sea level; *a c d e b* mass of the old survey craters; *c d e* cross-section of the former craters; *f* the valley of the Cumbre (Circus, or las Cañadas); *e* southern edge of the elevation crater (los Azulejos); *c f g* the volcanic cone of the Pic de Teide; *g* of its crater; *d h* the western crest at Pic, with crater *h* (Chahorra); *x, y, z* are parasitic small craters. [121]

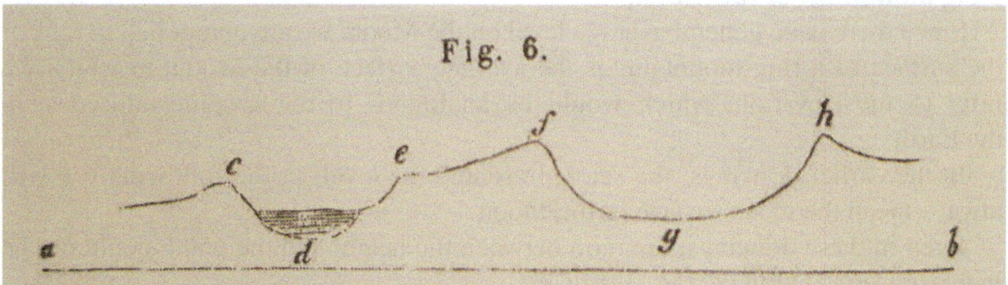

Fig. 6. Crater (Maare) of the Eifel; *a b* sea level; *c d e* and *f g h*, which are almost devoid of walls, one of which is partly filled with water. [122]

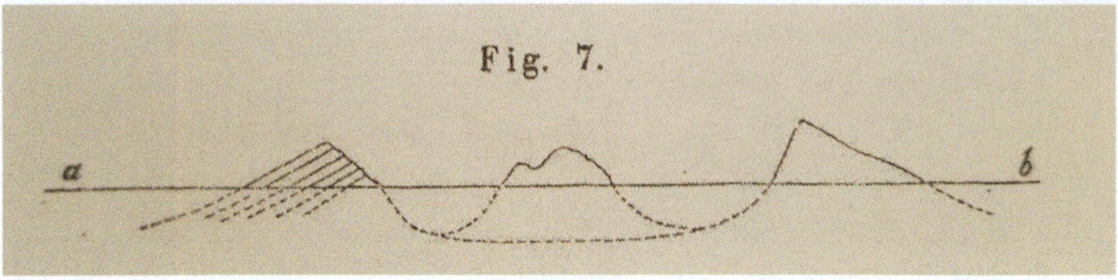

Fig. 7. A volcano rising from the sea.

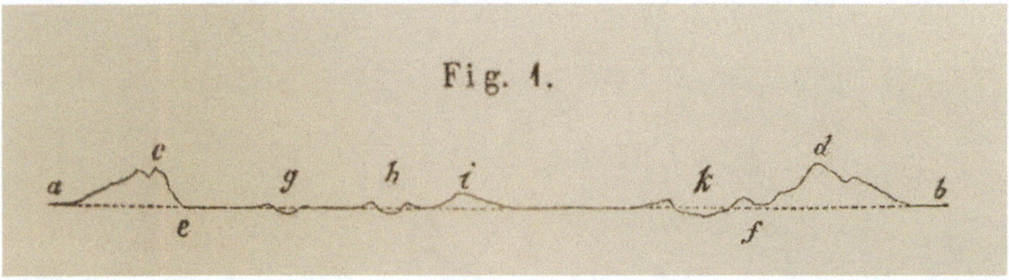

Fig. 1. Cross-section of a walled plain; *c d* ridge of the rampart; *e f* diameter of the inner surface, in the plain with *a b; g, h, k* are small craters; *i* is a hill. [124]

in reality, are explained in these drawings. Not readily do the slopes exceed 40°; often they are only 20° to 30°. At the Pico de Tenerife you can find in this respect the following conditions take place (see Fig. 5):
Gradient of the southern slope of the Pic cone in the direction g f = 33°;
Gradient of the inner valley f (from the foot of the pics down to the south) = 3½°;
Inclination of the inner wall of the southern rim of the old elevation crater, that is f e = 50°;
Slope from e to p = 21°, from p to q = 14° and from a to the sea at b = 4°. [123]

Ring mountains of the Moon.

Since there is no general average level on the Moon, we are compelled to look at the surface of a ring-mountain, as the average surface of the Moon, to which we refer to the elevation, which would be analogous to the absolute altitudes on the Earth.

In the earlier drawings, the sea represented the level; in the following *a b* will always mean the mean surface of the Moon.

Even in these drawings, the ratio between the height and the width could not be increased.-Fig, 1.-Fig, 2.-Fig. 3,-Fig. 4.

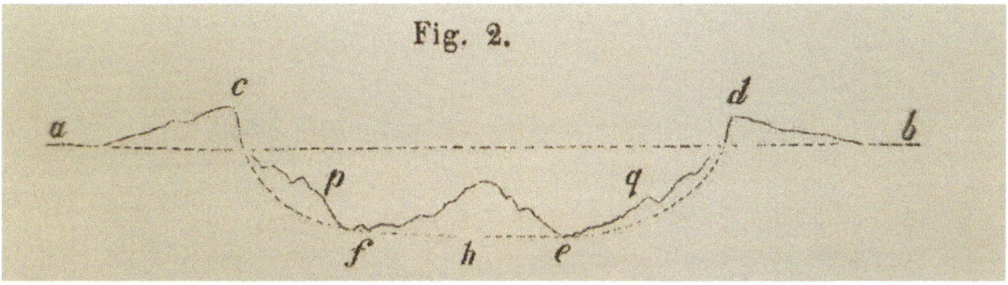

Fig. 2. Cross-section of a large crater; *a c* and *d b* outer surface of the crater wall; *c d* highest edge of the crater; *c h d* the inner slope of the crater, at the top of c and d 30° to 60° steep; *p* and *q* average of the terraces; *h* the Central mountains, whose summit lies under the average surface height of the Moon. [125]

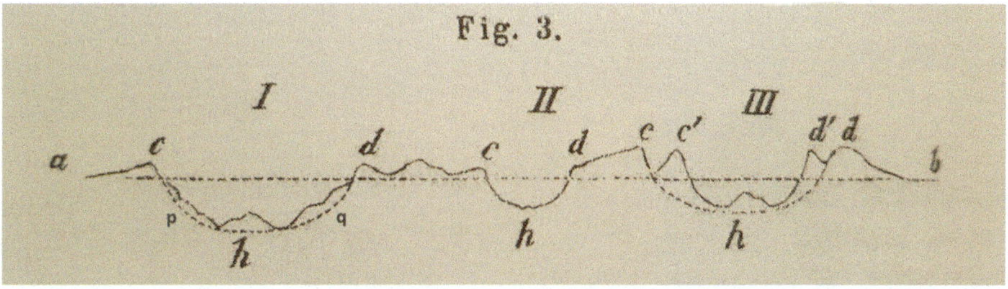

Fig. 3. Small crater; No. 1, a small crater whose rim lies in *c* and *d*, the terraces of which are in *p* and *q*, and the central mountain lie in *h*, No. 2 a still smaller crater, No. 3 a concentric double crater, *c h d* the eldest crater, *c' h d'* the later crater, whose Central mountain lies in *h*. [126]

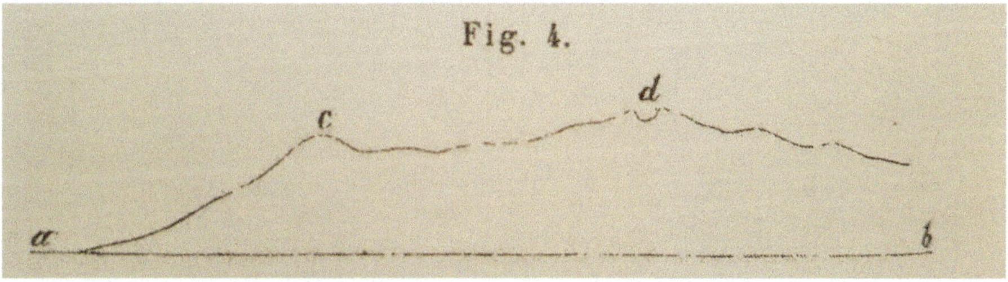

Fig. 4. Cross-section of the Huygens mountain range; *c* the Cap Huygens = 2400t high; *d* a small crater on the mountain Huygens, 2800t high.

Chapter XXIV
Dimensions of Some Craters of the Earth, Expressed in Toisen

With the exception of the altitude, we know little about the dimensions of depth and diameter. We can see from the notes where the numbers from the following table are taken.

	Absolute height	Largest diameter of the crater	Depth of the crater
Pico do Fogo [127]	1300t	4100t	140t
Caldera de Taburiente [128]	1193	3100	830
Caldera das Sete Cidades (Sao Miguel) [129]	435	1700	310
Caldera of Corvo [130]	400	1090	200
Caldera of Graciosa [131]	211	620	150
Caldera of Monte Brasil [132]	87	410	-
Caldera of Santa Bárbara [133]	550	1330	-
Caldera of Faial [134]	525	1050	-
Kilauea (Hawaii) [135]	608	2800	330
Tangkuban Perahu [136]	1000	-	130
Marapi	1400	1500	500
Mauna Loa [137]	2115	1800	200
Laach Lake [138]	244	1380	130
Rodderberg [139]	100	330	30
Vulcano [140]	204	500	100
Stromboli [141]	462	330	100
Mount Vesuvius [142]	600	350	100
Pico Alto [143]	1191	280	-
Mount Teide [144]	1910	100	20
Chahorra [145]	1547	280	22
Etna [146]	1700	250	54

(continued)

J. F. J. Schmidt, *The Moon*, Historical & Cultural Astronomy,
https://doi.org/10.1007/978-3-030-37269-9_24

	Absolute height	Largest diameter of the crater	Depth of the crater
Nevado de Toluca [147]	2370	500	192
Klyuchevskaya Sopka [148]	2465	370	-
Ruku Pichincha [149]	2490	769	380
Popocatépetl [150]	2771	830	333

Chapter XXV
Dimensions of Some Ring mountains of the Moon

As already mentioned, the absolute height of a crater is to be understood as the absolute height, and the depth as in the case of the craters of the Earth, i.e. measured from the rim to the base. The values are, as always, given in Toisen.

	Absolute height	Maximum diameter	Maximum depth
Theophilus	470t	50000t	2700t
Piccolomini	600	47000	2400
Copernicus	700	46000	2200
Tycho	-	45000	2500
Bulliadus	800	32000	1500
Scoresby	550	29700	1700
Burkhardt	720	29000	2500
Jacobi	500	27000	1600
Aristarchus	400	23000	1100
Manilius	500	21000	1300
Kepler	260	18000	1500
Theaetetus	550	15000	1160
Lambert	380	14000	900
Helicon	220	11000	960
Clavius *d*	450	10000	1500
Bessel	300	10000	700
Picard *A*	350	8400	1100
Picard *d*	300	5300	1200

J. F. J. Schmidt, *The Moon*, Historical & Cultural Astronomy,
https://doi.org/10.1007/978-3-030-37269-9_25

These figures have been determined, according to MÄDLER, partly by my own observations; they sometimes differ from some of the statements in Section XV, since the number of measurements has now increased.

Most of the crater diameters I approximately stated in Toisen, converted from MÄDLER's mileage; some were determined according to the maps.

If we determine the ratio of the depth to the diameter of the crater, the 18 ring-mountains give the following numbers:

depth from Nr.	1 = 1/18	of crater diameter.	Ring mountains between 50000 and 45000t diameter; ratio = 1/19
"	2 = 1/19	"	
"	3 = 1/20	"	
"	4 = 1/18	"	

depth from Nr.	5 = 1/21	of crater diameter.	Ring mountains between 32000 and 21000t diameter; ratio = 1/16
"	6 = 1/18	"	
"	7 = 1/12	"	
"	8 = 1/17	"	
"	9 = 1/21	"	
"	10= 1/16	"	

depth from Nr.	11 = 1/12	of crater diameter.	Ring mountains between 18000 and 11000t diameter; ratio = 1/13
"	12 = 1/12	"	
"	13 = 1/15	"	
"	14 = 1/11	"	

depth from Nr.	11 = 1/12	of crater diameter.	Ring mountains between 18000 and 11000t diameter; ratio = 1/13
"	12 = 1/12	"	
"	13 = 1/15	"	
"	14 = 1/11	"	

depth from Nr.	15 = 1/6	of crater diameter.	Ring mountains between 10000 and 5000t diameter; ratio = 1/8
"	16 = 1/14	"	
"	17 = 1/8	"	
"	18 = 1/4	"	

It may rightly be argued that these 18 values are not yet sufficient to undoubtedly prove a depth that increases relatively as the crater diameter decreases; however, it may be sufficient to remark that a 5 to 6 times larger number of measurements, regardless of their uncertainty, on average confirms the correctness of the above perception.

In general, it is much easier for the Moon than for the Earth to determine such circumstances.

Our knowledge of the shape and size of earthly craters is still very poor.

It seems that most travellers were more interested in the mineralogical investigation.

Not many like von BUCH, von HUMBOLDT, and ABICH, have tried to determine the general conditions, and still less were able to, instead of determining very uncertain estimates, to carry out accurate measurements under the greatest discomfort.

But even if one abstracts from the insufficiency of many observations of this kind, it is certain, that in the extraordinary diversity of the construction of earthly volcanoes, and especially in consideration of their intermittent activity, the depth of the craters is periodically subject to the greatest changes.

How little coincidence in the ratio of the depth of the craters to the diameter is easily seen from the following figures, which, though only incidentally, have ordered the nature of the volcanoes.

1. Caldera

I exclude Fogo for the reason expressed in Note 127

Caldera de Taburiente (Palma). .	= 1/4
Caldera das Sete Cidadas	= 1/5
Caldera of Corvo . .	= 1/5
Caldera of Graciosa	= 1/4
	Average = 1/4.5

2. Small active, simple or composite volcanoes; Sea level less than 1500t.

Vesuvius	= 1/5
Stromboli	= 1/4
Vulcano	= 1/2
Kilauea	= 1/5
Marapi	= 1/3
	Average = 1/4

3. Large active, simple, or composite volcanoes; Altitude higher than 1500t.

Mauna-Loa	= 1/9
Etna	= 1/5
Ruku Pichincha	= 1/2
Popocatépetl	= 1/3
	Average = 1/4.7

4. Presently inactive volcanoes.

Pic de Teide .	= 1/3
Chahorro	= 1/13
Nevado de Toluca	= 1/3
Rodderberg	= 1/10
Laach Lake	= 1/10
Average	= 1/8

The averages are illusory; where the individual indications differ so strongly, one has no right to ascribe to them any significance, and I only include them to show how much is still lacking in order to know whether or not there are certain dimensional ratios among the earthly volcanoes, whether they are connected with the nature of the elevated rocks, with the height of the sea, and with the peculiar qualities of the structure by which an old Caldera differs from the volcanoes which have risen in their midst, the latter again from the highest conifers over the ridge of the Cordillera, and from the rimless, often water-filled (explosive) craters of the Eifel.

Chapter XXVI
Closing Remarks

If we review the contents of all the sections devoted to the surface of the Moon, it will be easy to separate the purely scientifically investigated facts from the conjectures which have been made on the effects of nature on the Moon.

As results a result of the Moon observations we have so far, we may make the following statements:

1. We have a sufficiently complete knowledge of the mutual position and the spatial conditions of all the larger mountain forms on the facing hemisphere of the Moon.
2. We know the height of many mountains and the depth of many craters with sufficient accuracy.
3. The resemblance of the Moon-mountains with respect to their external form and grouping is very slight, compared with those of the Earth.
4. It is certain that no changes can be detected in the mountains of the Moon.
5. All the craters and most of the mountains were created by erupting forces; this is the most probable explanation, which follows directly from the observations; every other is compelled and arbitrarily presupposes circumstances that do not exist on the Moon; in the case of the assumption of erupting forces, in this case neither lava nor gas eruptions will necessarily be mentioned, and no distinction is made between elevation craters and ordinary volcanoes, which are necessary for the mountains of the Earth.
6. The Moon has absolutely no atmosphere on its visible half, and shows no phenomena which indicate with certainty such a very thin air.

The basis for a comprehensive selenography, as certainly the future will be, has been created by the works of LOHRMANN and MÄDLER; on the basis of this, it is less desirable to work out new generalities on a large scale, but rather to occupy oneself with the special observation of small surface areas of the Moon.

J. F. J. Schmidt, *The Moon*, Historical & Cultural Astronomy,
https://doi.org/10.1007/978-3-030-37269-9_26

May this work fall to those who are not dependent on the business of an observatory, and not on the vagaries of an unfavourable Northern climate.

May they be allowed to draw and measure the mountains of the Moon under the beautiful sky with more fortunate circumstances, to devote their remaining time to purely geological studies, to be free from scientific prejudices, the domain of knowledge unhindered by the theories of usefulness of one-sided naturalists and to understand the physics of the heavenly bodies in conjunction with that of our Earth from a common point of view.

Chapter XXVII
Opinions About Living Beings on the Moon and on the Planets [151]

It is not curiosity alone that encourages man to wander beyond the limits of knowledge; restless, and never fully satisfied in his search for knowledge, is not enough to look at countless miracles of the Earth, not the immeasurable wealth and the beauty of the forms and colors that glorify this world, not the understanding of great natural laws which finally follow him has become part of the vain efforts of many centuries.

Being accustomed, as a youth, to the sight of a part of creation, as long as he has not learned to observe and examine sharply, man is only too inclined to accept as well-known many phenomena which until today have eluded research; his attention is more aroused by rare and extraordinary, especially violent, effects of nature, than by events which, in his opinion, are entirely commensurate with the ordinary order and the never-changing course of events.

He is delighted to see the evening when the red dusk disappears on the horizon of the sea, and soon afterwards the glow of the sea, in the dark, the sandy coast, many of which are animated by animals; the turmoil of the thunderstorm and the subterranean thunder of earthquake threaten him; but he does not think of asking questions which are more to be answered by the spirit, by the scientific consideration, than by the statutes of popular physics, which can never separate itself from the external impression of the phenomenon.

The glittering glow of Sirius and the great stars of Orion unfold their quiet splendor when, for the first time, the shadows of the mountains of the Moon, or the rings of Saturn at the telescope; but more than the narrative of the results of the observation, more than the communications on the laws of the heavenly movements, he is interested in the questions which lie beyond the realms of modern science, which concern the inhabitants of the stars, and even for special purposes in the economy of the universe.

Every astronomer will have a hundredfold experiences that most people, from all states, prefer to see these questions answered.

J. F. J. Schmidt, *The Moon*, Historical & Cultural Astronomy,
https://doi.org/10.1007/978-3-030-37269-9_27

If we look for the cause, there are different means of explanation. The very slight spread of astronomical knowledge, and the lesser understanding of the subject of observing astronomy, prevent us from finding the correct point of view on such questions.

One overlooks too easily, or does not believe that astronomy has been working on its foundations for two millennia, and that the great telescopes of our time, preferably used to locate the stars, are much too weak to tell us about the particular nature of the many Millions of miles distant planets to provide sufficient information.

While the One in his opinion about the inhabitants of the Moon or the planets can rely on observational astronomy, the Second loses himself in this consequence, in order, merely to make philosophical considerations, to construct the world according to his ideas; the Third chooses a common hypothesis to fathom the purposes of creation on the grand scale, and the life of the beings on individual heavenly bodies, by means of his imagination, with the help of bold analogies, which give their origin derived solely from earthly phenomena, and therefore their limitation.

All these endeavors are based on something which differs from the demands of the still very frequent theories of usefulness, since they recognize not the independence and the higher profession of science, but only their practically useful purposes and successes; for every one will admit that little practical value would be to be boasted of those results which, in some way or other, have a convincing argument, For example, the existence of living and even rational beings on the Moon.

Now, in many cases, the question of the creatures on other heavenly bodies is but a kind of speculation, which has little firm ground, its reason is deeper in some people, where it is partly a purely poetic mode of presentation, or a more religious one. The eternal metamorphosis in animated and inanimate nature, to which man is subject, and no less an inborn impulse, which makes him think about the purpose and the ultimate goal of earthly metabolism, about his own destiny, and about the remaining of the Soul after death, easily directs the thoughts to the heavenly bodies; he arouses in a thousandfold forms, alternating between the feeling of earnest devotion and poetic certainty, a constant longing for what lies outside the Earth in a dark distance.

This longing has manifested itself in most great spirits of all centuries, as a striving after the earthly, whether it be in the sphere of religion, philosophy, or empirical scientific research; the spiritually advanced humanity has not given up such desire; it preserves it as an inheritance from the various generations, and is not tired of answering these and similar questions through science without wanting to undermine the rights of faith in this direction.

Depending on the viewpoint of knowledge, the size of the world view, and the depth and intimacy of the feelings. thinks and talks about the old man, who, after a thorough life, surveys all the results of his and foreign researches, differs from the one who, in the splendor of youth, preferred the great multi-faced dream full of beautiful and confused perceptions to the strict thinking and the troublesome penetration of knowledge.

So we all stand before the closed gate; there is no answer to our questions; monolithic only in the eternal stream, the light of the stars comes to us from the depths of space, and from very different periods of time it reports to the eye, states and orders in the universe, the interpretation of which can only be subordinated to the mathematical idea.

If, therefore, it seems as if we should forever give up the hope of obtaining in a direct manner through science the effect of nature on the surfaces of the planets, neither the legitimacy nor the interest of these questions is diminished Origin, as it seems, deeply rooted in the essence and in a need of the human mind.

It remains for us to designate the limits within which our wishes are to be restrained, and partly, if only negatively, can be answered by science.

Whoever does not disdain the sure results of observation and the sum of concurrent experiences, which we owe to the physical experiments and the observations at the telescope in conjunction with the calculation, has a right, as opposed to the dismissive tone, with sometimes the question of nature the Moon or the planet is answered, to follow a certain line of thought, which, if it is reasonable, at least has the advantage and the merit of eliminating a large number of erroneous views and ill-founded theories by further dissemination.

It is, moreover, who apprehends with more calmness, and in a more thorough manner, the results of astronomy, without abandoning himself solely to the empty astonishment of mere numerical expressions.

Large numbers in themselves, expressing unthinkable distances or masses, are unfruitful to the imagination; the less we are able to elevate ourselves to the right concepts, the less justified are the admiration which they excite.

He who seeks more in creation than the lawful order according to which the motions of the celestial bodies are guided, who does not merely turn his admiration to the acuteness of the human mind, is little satisfied by the greatness of space or the infinite.

It is not the numerical expression alone that dignifies the sublime character that reveals the majesty of the works of nature.

For this reason, if we are conscious of the limits of the infinite in the futile endeavor to grasp the infinite, we return to the contemplation of organic life, and to the freer view, according to which we enliven the planets and assume creatures everywhere in the universe, who want to be called to higher purposes.

So far, there is no obstacle to contemplation. Neither reason nor the standpoint of science can rise in a positive way against it.

But the question now arises, of what kind of constitution may the creatures on other heavenly bodies have?

To answer this, it is necessary, if we do not wish to be completely left to the imagination, to first of all ask all the necessary questions:

"What do we know today about the surfaces of the Moon, the Sun, or the planets, and of the possible conditions of life that make organic beings possible there?"

This question, upon which everything depends, we will try to answer in a few words.

For more than 200 years the telescope has already been used for celestial observations.

The instruments have gradually improved, and according to the magnitude and the optical power of many of them, many have promised extraordinary successes.

But, apart from these overly exaggerated expectations, the question of the organisms of other celestial bodies all too often overlooks one essential circumstance, Gravitation, which must first be considered, and more important than the possible existence of atmospheric air.

According to the astronomical calculations, as one of its most reliable results, is that by setting the gravitational force = 1 on the Earth's surface, the value on the Moon is 1/6, and on the Sun the value = 28, how this great difference of gravitation gave the first and most important reason for the conviction that living Beings on the Sun or on the Moon must be quite different from those on Earth.

For these differences of gravity are not limited in their effect to the fact that on the Earth a solid body travels 15 feet, on the Moon 2 1/2 feet, and on the Sun 428 feet free-falling in the first second, but they affect every organism with respect to the ability to move, to increase through growth, or to apply any powers.

Our body would, on the Moon, be found to have astonishing effects with the greatest ease, because the heaviness would oppose it with a slight lower resistance, but it would be on the Sun, under the influence of a 28 times greater severity, not only be hindered in all movements and activities, but also, apart from quite other causes, to wither in no time. [152]

Since, on the bodies of the solar system, gravitation is more or less different from that which we find, we must concede, for this reason alone, that although there are beings of all possible forms and abilities, not one of them can exactly compare with terrestrial organisms.

If we examine the question of the physical nature of the surfaces of various celestial bodies, and if we limit ourselves to the planetary system, we shall find that we have on the whole received very little sufficient information through direct observations, and that for our purposes no result of positive evidence, which would be able to determine the existence of living beings, let alone their characteristics.

The telescopic observation teaches that the light-radiating surface, the Sun's photosphere, is very often covered with significant groups of patches, which show a maximum of their frequency from 11 to 11 years.

These irregularly rounded black-grey shapes are funnel-shaped depressions in the Sun's light envelope, through which we evaluate completely or you can look down on the dark body of the Sun.

We distinguish the black nucleus, and the gray penumbra surrounding it, i.e. the edge of the corona, whose edge is either sharply defined or faded, transitions into the photosphere, covered with countless small grey pores or scaly like dots.

Often such patches are 5,000 to 10,000 miles in diameter, and groups have been observed which, composed of several hundred patches and dots, covered spaces of 50,000 and 60,000 miles. Moreover, in certain parts of the Sun-disc, especially at its margin, Sun spots are noticed, the brightness of which is greater than the ordinary light of the surface; they are called flares or light-clouds, and these, as frequent companions of the black spots, form perhaps flat, bulging, but transitory elevations.

Probably at all times light gaseous matter hovering over the surface of the Sun, flowing with great force out of the blotches or near them, in total solar eclipses within the form of ruby-red flames or mountain-like phenomena of 100 to 10,000 miles.

Moreover, it is known that the Sun is rotating in twenty-five and a half days; we do not know any more, and no one will be able to believe on the basis of these perceptions that on the surface of the Sun, which has been radiated with dazzling light and perhaps extraordinarily heated, there are living beings similar to those on Earth under the effect of a force 28 times greater.

We also know very little about the nature of the planets Mercury and Venus.

Certainly, these bodies have suggested atmospheres, such as the attenuation of light at their phases, or a rare ephemeral formation of spots; they also have mountains which are manifested by the irregularities of the phase or the "horns"; they both probably turn around their axis in the approximate time of a day, and are not very significantly different from Earth in terms of gravity at their surfaces.

But on them the sunlight is stronger than with us; up to 10 times the daylight on Mercury surpasses ours.

An organic nature appropriate to the planet Mercury and Venus may possibly be somewhat similar in some respects to earthly ones, but may not match it.

The planet Mars is considerably smaller than the Earth, and the gravity on its surface reaches only half of the amount of ours. But observation with large telescopes shows us phenomena, which sometimes resemble very similar ones on Earth.

Variable dark spots, diminution of light at the edges, uneven intensities of the disc's yellow and orange-red light make a strong atmosphere very likely.

There is reason to suppose that in the patches we see partly clouds, partly oceanic surfaces.

The poles of Mars are white-luminescent, and their regions often distinct and circular from the dim yellow-red light of the planet; these white, very noticeable areas enlarge and diminish as the pole in question has winter or summer, and there can be no doubt that they derive from atmospheric precipitation analogous to our snow, the distribution of which depends on the seasons. [153]

Mars rotates in 24 hours 37 minutes and the intensity of its illumination by the Sun is only half of that which the Earth receives. Its mean distance from the Sun is 1 ½ when our planet is set = 1. All in all, it is justifiable to conclude that if animals and plants live on Mars they may possibly have some resemblance to terrestrial creatures, because the planetary conditions there, as far as we know them, do not directly contradict this inference. A complete equality is unthinkable. If nature were so poor and so limited to the earthly forms as some predestined in their speculations, we would have to believe that on the planet Mars the animals and plants there are in a very miserable state of development the atmosphere is identical with ours, and organic creation is but an imitation of earthly types with similar bodies and functions.

Instead of pursuing such unworthy ideas, it is better to remind oneself of MÄDLER's statement: "Nature does not like to copy itself; it is rich enough to create individuals, yet it preserves unity in diversity."

Of the numerous asteroids between Mars and Jupiter, we only know the orbits and only know that these celestial bodies are very small.

Some may be barely 12 miles in diameter; in any case, the gravity of their surfaces is extremely low.

This circumstance, and the considerable distance of the asteroids from the Sun, completely excludes a similarity of their possible organisms with those of the Earth.

Jupiter is 11 times larger in diameter than the Earth; it rotates in 9 hours 56 minutes; its days and nights are thus only about 5 hours long, and the former have brightness 25 times less than the days of the Earth.

At the telescope, we notice phenomena which indicate a strong atmosphere, and the numerous grey stripes, parallel to the equator, which often dissolve into individual rounded spots and dots, or at least show considerable compaction, may at least be regarded as analogues of our clouds. But the gravity of Jupiter's surface is 2 ½ times greater than with us. [154] This fact, the great distance from the Sun, the brevity of the days and nights, suggest that the atmosphere and the structures of possible creatures are very different from those of our Earth, and that we must confess that no attempt is made to create creatures of other celestial bodies, so to speak, by imagination, no matter how ingenious and cautious, could lead to any satisfying goals.

We would arrive at the same conclusions if we wanted to take into account the distant Suns and the great planets Saturn, Uranus, and Neptune; it suffices to recall that conditions of life, as we know and regard it as necessary on our Earth, are to be presumed on these three bodies to a much lesser extent than on the already discussed planets from Mercury to Jupiter, and it will be completely unnecessary to give special consideration to the long-curled, very transparent, almost bodyless comets moving in very eccentric orbits.

Only the Moon, so close to us, whose mountains we can see so clearly and survey with such certainty, still seems to offer some hope.

But even this must soon disappear, if one summarizes the facts already mentioned in these pages.

Nevertheless, the Moon may nourish innumerable creatures, even animal or plant-like, and also rational beings; only to give up the hope of seeing them, and to let the opinion that organic nature on the hemisphere of the Moon on this side is in any way comparable with the Earth.

In order to perceive larger creatures of the Moon as distinct but almost imperceptible points, and more importantly, to recognize them in their motion, we require a magnification of about one hundred thousand times magnification on otherwise perfect telescopes; and a perfectly accurate movement, which moves the instrument so as to follow the apparent motion of the Moon in the sky, and, as a principal requirement, an absolutely transparent and motionless air, which allows the application of a hundred thousand magnifications.

But now, even on our most excellent telescopes, we can only use magnifications of 300 to 1000 magnifications in the dire Central European climate, with very little benefit for the Moon; the former, according to my experience, perhaps three or four times a year, the latter never, provided that the highest demands are placed on the purity and calm of the optical image, as is absolutely necessary in the finest investigations of very small objects on the lunar surface.

While it is true that in some places in the tropical zone the purity and stillness of the air is much better than in our Nordic countries, it seems undoubtedly that nowhere on Earth, excepting the highest peaks, the nature of the air is ever better would allow for the advantageous application of a hundred thousand magnification. [155]

It is, however, possible that we may at one time discover objects on the Moon, which are perhaps not to be attributed to the works of nature, but to the activity of living beings: this will educate the future.

To this day we know absolutely nothing about it; if dreamers have seen what they wished to see, then their miraculous views belong to the realm of the fables, which, of course, is very remote from entering or extending this work.

If one considers recently that the gravity on the surface of the Moon is 6 times less than on the Earth, that its days and nights each last 354 hours, that it lacks any verifiable trace of an atmosphere, it must be reasonably conceded that the whole arrangement of nature on the Moon is entirely different from earthly nature in all respects.

There are many and varied reasons for the question of the habitability of celestial bodies in general.

The latter cannot answer science directly; it can only prove where conditions exist that allow earthlike organisms or where they are absent.

But the question of whether humans, animals and plants live on the Moon or on the planets is already a misguided one, because in most cases it is not even considered to inquire into the more or less well-known natural conditions on those bodies.

It is a very fruitless enterprise to enumerate the earthly organisms which thrive under the least atmospheric pressure, in the coldest temperature, in the total lack of light, in order to signify through them, as it were, the transitions to the beings on the airless Moon: that is, human limitations to the works of nature and its creations,

to enliven all vitalizing power, to give her childish advice, how she could help herself, where living conditions seem to be lacking, which we rightly regard as necessary for our existence.

It is also not to be condoned to subject certain principles to the utility and necessity of nature.

If any earthly organism requires certain means for its development and for its purposes, it is not necessary that the same agents be ascribed to similar organisms on other worlds.

One means of communicating with the outside world is vision; that inhabitants of the Moon must see, is not to be considered necessary even according to earthly concepts, much less that they must possess certain more perfect organs, because they, according to human conception, would require them.

If we make analogies, and always carefully compare them with physical or astronomical facts, we are, at least to the best of our ability, assured against the transgressions of the imagination, the assumption of uncritical fictions; but we do not push much further.

The rich creation on our Earth is after all only an isolated phase of the entire natural life; we come to this view, in order to rid ourselves of the age-old childlike delusion, according to which the Earth is regarded as the centre of the universe; but we gain only a greater far-sightedness, by no means any insight into the general plan according to which organic nature presents its countless forms under such diverse conditions of life.

We are like the traveler who looks up from the plain with desire to the towering mountain peak.

Once at the top, he looks over the fields and over the army of floating clouds, to the distant edge of the horizon, where the blue of the sky blends with the line of the sea in a bluish scent.

The farther he looks into the distance, the dimmer the pictures become.

Just as, according to Mädler's expression, the bodies of the world are not copies, but individuals, one would think that this closed individuality also refers to the organisms of the individual heavenly bodies, that these forms of life are not replicas or metamorphoses modified by planetary conditions, but free creations, peculiar, and appropriate only to the heavenly body that they occupy.

But it also appears as if the primitive character of living nature had changed within the sphere of a heavenly body, and to a certain degree dependent on the catastrophes which its surface had undergone during the course of immeasurable time periods.

In ancient times, many animals, completely unknown to now, lived on Earth under the influence of enormous reptiles, monsters of extraordinary shapes; wonderful trees and shrubs grew, alive with insects, at the foot of old, now extinct volcanoes.

Everything has disappeared; the fossilized remains of those creatures give us the only but unequivocal indication of their former existence. [156]

The later world shows none of the old animals and plant forms; new ones have arisen, which in no case can be regarded as the product of the generation of the former creatures.

Here we are confronted with the darkest problem, but we are aware of the power perfection of nature, which can perpetually change the character of a planet's organisms without resorting to the usual metamorphoses known to us; from this repeated act of creation on our Earth, we infer similar ones in other regions of the heavens.

But wherever the gaze turns, it will never quite be satisfied; man is not allowed to guess the beginnings of the created, the dark hieroglyphics, which mysteriously conceal the working of matter, and conceal in it and through it the ultimate and highest principle of the universe.

May every one of them, whether he is self-reliantly called to foster science, or only enjoy the fruits, the results of which, and rejoice in their vigorous, vital progress, content themselves with that which is attained today, hoping for a greater future; which is imminent to the aspirations of man; it praises itself happily the one who wishes and may hope that our spiritual pursuit of the end of this existence has yet to be traversed by unknown paths.

Chapter XXVIII
A Day and a Night on the Moon

When we think of the Moon in our minds, and choose a favourable point of view, it will not be difficult, in the light of known facts, to give a clear picture of the scenes of nature on our neighbour's world.

The place of our observation is the summit of the Central mountains in one of the larger craters, which, near the equator, is a little distant from the centre of the visible hemisphere of the Moon.

There it is night at the time when the new Moon appears to the Earth.

Almost at the zenith the full, two-degree diameter disk of the Earth shines; it emits a thirteen times greater quantity of light into the Moon-night, than when the full Moon shines towards the Earth. [158]

Over the course of several hours she soon shows the bough-rich continents of the eastern hemisphere, bounded by the dark surfaces of the ocean; [159], and in white splendour the pole, covered by the eternal snow and ice, which has just been turned towards the eye. Gradually, the eastern continents of the Earth disappear with increasing shortening, and in the middle of the light-reduced Earth disk [160] we see the Atlantic Ocean, west bounded by the still much shortened brighter border of the American mainland.

Thus the rotation of the Earth gradually brings to light most countries and seas. In the meantime the great night-time celestial body has scarcely moved away from the zenith; and as the stars of the Zodiac pass by it, the diminution of its circular disk begins, the visibility of the phase, markedly clouded by the atmosphere of light.

The stars shine down to the horizon in the unaffected splendour; by the appearance of the Earth neither the Milky Way nor the visibility of the small stars of the 6th magnitude is extinguished.

The Constellations [161] are complete, the order of the planets almost the same as they are seen from the Earth, only the apparent rotation of the sky does not take place around the world pole of the Earth, but around a point in the Constellation of the Dragon. [162]

J. F. J. Schmidt, *The Moon*, Historical & Cultural Astronomy,
https://doi.org/10.1007/978-3-030-37269-9_28

Twenty Nine times more slowly than for astronomers on the Earth, a star at the equator of the sky seems to be moving in the telescope by virtue of the daily rotation of the *primum mobile*. [163]

Around us the landscape is brightly illuminated by the light at the zenith of heaven; there is no shadow; we recognize the base of the Central mountain, the crater floors and the wall mountains, the proximity and the distance in the same clearness, no haze and mist obscure the purity of the firmament, we never see colourful edges or optical halos around the disc of the Earth; the night neither illuminates the redness of the northern lights, nor the sudden glow of lightning.

We note the slow course of the night by the ascending of the stars in the east, by the decline of others in the west; we can recognize it more closely by the more and more waning Earth.

For the time being, we see them completing their upheaval, until they have taken off the circular shape to that of the half Moon, the last quarter of the Earth.

Seven times we see her completing her revolution until she has changed from circular to half-Moon, the last quarter of the Earth. Its dark half shines with a gleam: an ash-coloured reflection of the moonlight. [164]

Already the coming of the day is at hand, but no dawn, no paling of the stars announces it.

It is only in the east that the white light of the zodiacal light has developed vertically in a high narrow triangular form, a representative of twilight at a time when darkness has won the night because of the dim glow of the Earth. [165]

In the east, Venus appears as a morning star casting shadows; in the depth of the craters towards the west, the mighty shadow of the Central mountain is evident in all its splendour. [166]

Occasionally we see in the sky the lonely meteor with a white, now extinct trace of the tail. [167]

In vain, we look for the signs of the day in the direction eastward; neither on the distant horizon, as far as it is known through the gorges [168], nor above the mountains, hovering in the morning, hangs red-rimmed cirrus clouds.

Suddenly, there are small but incredibly bright lights to the west; in a few minutes they outshine the brightest stars in the direction towards evening; they are the highest peaks of the western crater walls, which are hit by the first rays of the uppermost edge of the Sun above the east. [169]

The luminous peaks increase, become united with narrow, wave-like, light-emitting seams, and soon the whole profile is fully developed; but for the sake of contrasts it becomes impossible for us to recognize its connection with the base of the mountain, which could be seen clearly beforehand in the light of the Earth.

Spanning half the circumference of the western horizon, the upper rim of the craters, illuminated by the rays of the rising Sun, seems to float in the dark star-strewn sky.

Now we see above the mountains in the east, and at the base of the zodiacal light, a narrow white seam, the upper part of a small circle; we see the nimbus surrounding the Sun, which for the Earth-inhabitant at the time of total solar eclipses around the dark Moon in the form of the Corona occurs.

It rapidly increases in breadth and brightness, a last messenger of the new day; the highest point of the solar edge rising for the central mountain follows, and without a transition the full day seems to have subsided in a few seconds. [170]

After about an hour the whole Sun-disc illuminates the summit of our mountain, and its pointed shadow projected in the west against the terraces of the crater wall. Around us in the depths there is an impenetrable night; towards the east, every trace of the mountains has disappeared. [171]

In the black sky the brighter stars shine as before; the zodiacal light has been extinguished, the Earth has assumed the shape of the sickle with a concave phase.

In the immense contrasts scarcely tolerable to our eyes between the immeasurable quantity of light which the western mountains reflect, and the perfect darkness of the shaded depth, we feel isolated as if resting in a balloon; without support, the shining peak of our Central mountain seems to float freely in space.

The higher the Sun rises, the more the details of the surrounding landscape are revealed. Already all the western terraces are illuminated, their narrow gorges filled with absolute dark shadows.

We recognize craters of the smallest kind among them, and at the base of the terraces the crests of the low-lying hills gradually emerge as brightly radiating surfaces. [172]

In the east the crater wall is only negatively visible; from the north through the east to the south are all the rising stars obscured by an irregularly undulating narrow zone. Only in the East may their visibility be hindered by the Sun. [173]

Soon daylight also penetrates into the middle of the crater depth. In the west the whole landscape is visible, with the exception of the details in the deep valleys and in the space covered by the long, long, conical shape of the shadow of the central mountain.

Thus, in the early morning, the mountains emerge almost suddenly from the night, not from grey twilight, nor from the steaming misty valley, but without passing from the deepest darkness of the shadows of the mountains thrown from the east.

During this transition from the long night to the day, in this morning scene in a strange world, no sound reaches our ear. [174]

The new day does not awaken the familiar voices of the animal world, and there is no wind in the thickets of the trees. No bird flies to the black sky; the desolate soil does not adorn a plant, nor any known animal.

What covers it in wonderful forms, in forms of life, cannot be deduced and understood by our knowledge. It is silent on the ground, as in the sky, with the Sun, the crescent-shaped Earth, and many stars glowing on the dark vault of the cloudless purity.

In vain does the eye look for shining sea-surfaces, or between the ravines of the mountains, the vastness of the dark sea, in vain for the enchantment of the landscape, on the Earth soon by the colours and the abundance of vegetation, by light and air, by clouds, and cloud shadow, or the grouping of snow-capped peaks over forest-covered slopes of the mountains, the degree of our joy, and our particular interest.

We may not hope to find anything new at the bottom of the crater.

We feel the descent much less the effect of weight, and to the extent that effort and fatigue are diminished, the fear of danger and dread also vanishes at the sight of near close abyss on the edge of steep rocky walls.

We see monstrous blocks giving way before our usual exertion; There is no roar following its fall, no echoes from the mountains.

We reach the depth where no fire is burning and no lava flows; here, too, we seek in vain for known forms which are comprehensible to our ideas.

Where the ground is flat and strikingly dark, where the reflection of the rays of the Sun increases the warmth of the neighbouring mountains, we are never surprised by an earthly plant form; neither tall palm trees nor gloomy aloes enliven the dry light-radiant rock.

What we see near or far in motion, individual bodies or groups of them, remains incomprehensible, and because of inequality of the sensory organs, we lack every means of making ourselves understood, and to attract their attention from afar. [175]

Under such considerations, on that day, on the Moon, many hours of the long mornings may pass. The midday Sun is approaching; the Sun's rays pass through the zenith, and through the meridian.

Very close to the Sun, only the finest sickle of the Earth can be seen with difficulty.

It will take another few hours to wait for the dark body of the Earth to appear before the Sun, finally causing the great spectacle of a total eclipse.

In the place where both discs are touching, the edge of the Sun becomes turbid, and soon we see the flatly curved edge of the dark Earth, which, in constant progress, diminishes the light of the day.

In the course of an hour only a short crescent-shaped piece has remained of the Sun: [176] it rapidly decreases in size, and before this disappears, we see westward in the direction which the shadow cone of the Earth follows, all mountains obscure themselves through contrast. [177]

With the disappearance of the last ray of sunshine, the deepest night has come; more than ever, the vault of heaven is beset with innumerable non-shimmering stars. [178]

At the zenith, at noon, the Sun is gone; the powerful, negatively visible, black circular shape of the Earth is surrounded by a broad and shining gleam of light, caused by the atmosphere and the outermost light-enveloping of the Sun, whose fiery glow illuminates the mountains with a reddened glow like a winter landscape in the reflections of the colourful Aurora. [179]

In the course of an hour the intensity of the nimbus surrounding the Earth gradually changes, the stars of which are easily recognized in the near vicinity [180], and soon the greater splendour betrays the place where the first ray of the Sun will shine again.

As we await this moment, the peaks of the mountains glimmer with a bluish light [181] in the far west, and after a few seconds the darkness and splendour of the drama have come to an end.

Slowly the shimmering hemisphere disappears around the Earth, [182] and in the east the mountains are revealed by the veil of shadow; the smallest stars disappear, and after the Sun has regained its circular shape, we see, a few hours later, the fine, increasing crescent shape of the Earth.

In the seven-day afternoon we see the Sun turning more and more to the west from the zenith; all around us black spots appear in the landscape, the first short shadows, and the dazzling uniformity of our surroundings soon attains the strange appearance of the morning landscape, in which still illuminated masses of mountains appear more and more separated like islands by the complete darkness of the shadows.

The whole of the western ramparts dissolves into glittering surfaces, and only the upper hemispheres of the highest terrace still radiate in the form of an irregularly curved gold strip, which slowly separates into individual pieces, finally disappears in numerous points of light.

The shadow of the western wall reaches as far as the middle of the crater, and the shadow of the central mountain begins to rise on the eastern terraces.

At the moment when our view of the top of the mountain, the upper edge of the Sun goes down west of the mountains, we are surrounded by night, and see neither the mass which carries us, nor the depth from which it ascends; the entire eastern horizon encompasses the illuminated upper crater wall; the more it decreases in breadth, the more it disintegrates into individual glittering spots; in the east, only the highest peaks, like great stars, flamed; they also disappear, slowly diminishing in size and brightness.

The night has begun; the Earth is again half-illuminated, and the host of the stars is spread over us in full glory.

Such is the dream of a day and a night on the Moon, from which one awakens one feels that his simple and dignified presentation is not without uplifting inspiration and some instruction.

But fairy tales and fables of the Moon tell another.

Annotations

1. The Moon, according to its cosmic and individual circumstances, or general comparative sellenography, with particular reference to the authors' publications of the "*Mappa Selenographica*" by W. BEER and Dr. J. H. MÄDLER. Berlin, 1837.

2. MÄDLER loc. cit. p.1 to 24 contains a common explanation of the lunar motion, which can also be read in the "*Populären Astronomie*" by the author, or in many similar works.

3. This table is taken from the MÄDLER works, where it is on p. 48 more complete, from +10 to −10 seconds of parallax. A geographical mile is equal to the fifteenth part of an equatorial degree of the Earth.

1 geographic mile	= 22842.55 Parisian feet,
	= 3807.09 Toisen,
1 Toise	= 6 Parisien feet,
	= 1.94904 Metres,
	= 6.39459 English feet,
	= 6.21002 Rheinland feet,
	= 0.97312 Vienna Fathoms.

4. *On new tables of the Moons parallax* by J. C. ADAMS M.A. in *the Nautical Almanac* 1856 and in ENCKE's "*Astronomisches Jahrbuch*", 1856 p. 301 IV. Included with ADAMS words p. 305:

   From the value of the nutation constant of Mr. PETERS, the value of the ratio of the mass of the Moon to the mass of the Earth is 1: 81.5.

   If we apply this relation, and connect it with the dimensions of the Earth according to BESSEL, and the length of the second pendulum under the state of

J. F. J. Schmidt, *The Moon*, Historical & Cultural Astronomy,
https://doi.org/10.1007/978-3-030-37269-9

35 1/4, which is derived from the FORSTER pendulum experiments in BAILY's account, I find the value of the Constant of Parallax = 3422" 325.

HENDERSON, however, in the essays (10th volume of the Memoirs of the Royal Astronomical Society) found the value of this constant from the comparison with DAMOISEAUS tables to be 3422" 46.

However, it must be noted that what is called parallax in the tables is strictly speaking the sine of the parallax, in seconds.

HENDERSON has assumed in his calculations that the values in the tables represent the parallax itself, so that the value found must be reduced by 0'15 in order to obtain the constant of the sine of the parallax.

The value derived from this is then 3422"31, a result which agrees with the theory in a truly admirable way.

— Hence the two values of the parallax, which are the same as the theory and the observation, are only different by 0"015, and if we wish to consider this difference as the remaining minimal uncertainty of the parallax, the mean distance of the Moon from the Earth is only 0.26265 miles, or 862 feet, a magnitude much less than the height of many of the lunar mountains.

According to the average value of the parallax = 57' 02" 32, the average distance of the Moon is very close to 51800 miles.

5. Comparisons can be found with MÄDLER. loc. cit. p. 5

6. The mean apparent radius of the Moon is inferred from Burckhardt = 15' 31"95 from meridian observations, from FERRER = 15' 31"68 derived from solar eclipses and star occultations.

   From his Königsberg heliometer surveys, Dr. M. WICHMANN an increase of BURKHARDT's Moon radius of + 1"06, the radius therefore = 15' 33" 01.

   This information is found in the important treatise WICHMANNS on the physical libration in SCHUMACHER's *Astronomische Nachrichten* Nos. 619, 621, 628, 630 and 631.

7. These figures are taken from a complete, but only slightly calculated, table of TOBIAS MAYER's: *Kosmosgraphische Nachrichten* and collections on the year 1749. Section IV.

8. The apparent enlargement of the Sun and of the Moon seems to be due to deception only so much as the fact that we estimate very distant, especially steep mountains, higher than they are. Through the effect of the air, one believes the mountain further, and in the case of an inadequate idea of its true dimensions, we also surmise it.

   —When I first saw the high peaks of the high mountains from Bern to the south, I felt deceived by the magnificence of the sight, that I could ascribe elevations of at least 5 to 6 degrees to the highest mountains.

   If I use the altitude of Bern above sea level = near 300 toisen, the relative heights of some snowy peaks according to G. STUDER, and the distances of them from Bern using a larger map, I find the following angles (ignoring the effects of refraction):

Apparent mountain heights seen above the horizon of Bern:

Jungfrau	= 3° 27
Eiger	= 3° 24
Finsteraarhorn	= 3° 06
Schreckhorn	= 3° 11
Wetterhorn	= 2° 56

The peaks of the Tatra mountains seemed to me to be extremely impressive on the Lissa Hora [Lysá hora] in the Beskids. The calculation shows, however, that even the elevation of the top of the Lomnic peak, about 1500 Toisen, was only about 35 arcminutes above my horizon. — Comp. with the work of A.V. HUMBOLDTS on the elevation of the Tolima of S. Fe de Bogota: *Kleinere Schriften*, vol. 1. p.108 and 109.

9. More specifically, the Moon mass determined by LINDENAU = 1/87.73 - According to PETERS investigations it follows = 1/81.5

10. If we set the mass of the Moon = 1/81.5, the mean density of the Moon = 0.60429, when the Earth = 1.

The density of the total mass of the Earth has been sought by means of various methods. We find an overview in the excellent geognosy of NAUMANN, vol. 1. p. 32 et. Seq. A more detailed explanation cannot be given here, and only the numbers which are known up to now are given in the following summary:

If the average density of the pure water = 1, then the mass of the Earth is:

= 4.713 according to the observations of MASQUELYNE and HUTTON 1774 to 1776 on the mountain Shehallion in Perthshire. (Calculation of ED, SCHMIDT: *Textbook of Maths and Physics*, Vol. II p.479

= 4.837 after CARLINIS Pendulum observations on the Mont Cenis 1824, also recalculated by E. SCHMIDT.

= 5.52 according to CAVENDISH's observations in 1797 with the torsion balance.

= 5.4383 according to REICH's observations with the torsion balance 1837 to Freiberg. (NAUMANS Vol. I, p. 38 and Cosmos, Vol. III, p.446).

= 5.67 according to BAILEY's investigations.

Assuming, as happens by many, the number of the most accurate is 5.4383, and if we hold the mass of the Moon = 1/81.5:

Density of the Earth	= 1.00000 or	= 5.44
the Moon	= 0.60429	= 3.28

(When the density of the pure water is taken as a unit.)

The mean density on the surface of the Earth can be assumed to be half of the general density of the Earth (NAUMANN, vol. I p.39) is assumed to be =2.5, half of the general density of the Earth; If this ratio is applied to the Moon, the surface of the latter is about 1.5.

Matter of the density close to 2.5 (those of the water = 1) are, for example,
Rock Crystal (2.65),
magnesia (2.30),
white glass (2.30),
Gipsum (2.32),
Potassium (2.66),
Carrara marble (2.73),
Natron (2.80),
Sandstone (2.40), etc.

The density = 1.5 is greater than that of amber, oak wood etc., but smaller than that of ivory, graphite, sulfur, etc.

By applying the mass of the Moon 1/81.5, the gravity on the surface of the Moon = 0.16528 = 1/6.05 is the same as that which causes a solid body in the airless space on the surface of the Earth, falls 15.11 Parisien feet in the first second.

For the Moon, this fall height is therefore = 2.49 par. Feet. (The number in the text refers to the mass of the Moon = 1/88). MÄDLER rightly remarks that for the relations of human forces, if these could be affected in all respects unchanged on the Moon, all the movements carried out in the horizontal and vertical direction would be 6 times easier and safer than on Earth.

11. The direct observation shows no flattening of the Moon. All detailed observations since T. MAYER lead to the conclusion that a recognizable flattening does not take place. This can be seen from the three series of observations which follow, of which I refer to the former only in order to show what consistency an excellent astronomer must be content with more than a hundred years ago.

MAYER micrometrically measures the diameter at the time of the full Moon; he found, For example, (Kosmogr. Nachr. V. Abschnitt) the deviations from the mean result:

in	the	1st	Observation =	−1"
"	"	2nd	"	+12
"	"	3rd	"	+6
"	"	4th	"	+7
"	"	5th	"	−3
"	"	6th	"	+11
"	"	7th	"	−11

He concluded from this by considering these deviations as observational errors that flattening was not perceptible.

BESSEL, with the Koenigsberger Heliometer, twice the Moon's diameter at the time of total darkness, if no perceptible phase takes place. He measures in different directions.

The deviations of the individual observations from the mean, ordered according to the positional angles at which the diameters were observed, are as follows: (*Astr. Nachrichten*, No. 263 and MÄDLER, p.10 Note.)

				1830 Sept. 2.	1833 Dec. 26.
with	0°	deviation	=	+ 0" 46	+ 0" 09
	30			– 0" 08	+ 0" 05
	60			+ 0" 14	+0" 10
	90			– 0" 41	– 0" 22
	120			– 0" 21	+0" 15
	150			+ 0" 12	– 0" 16

BESSEL concluded from these deviations that the circumference of the Moon must be very close to circular.

By way of theory, however, one has come to the conclusion that the Moon, in double respect, must show a slight deviation from the pure spherical shape, namely:

1. a polar flattening due to the very slow rotation;
2. an equatorial swelling of the hemisphere, caused by the gravitation of the Earth.

TOB. MAYER already dealt with the determination of the anomalies, and gave his result in a thorough treatise in the 3rd Section of the *Kosmogr. Nachrichten* for the year 1748.

By setting the ratio of the densities of the Earth and the Moon = 4:5 according to NEWTON, he finds that the axis of rotation of the Moon is equal to the diameter of its equator as 215770: 215771, or the equator is 1/215770 greater, i.e. about 4.1 toisen.

But since the density of the Moon is less than NEWTON assumed, we find the polar flattening by assuming the density of the Earth and the Moon = 5.46 and 3.07 assuming 9.2 toisen.

According to this, it is impossible to measure flattening by practical means, since from the average distance of the Moon, 4.1 toisen appear to be only 0 "0043 and 9.2 toisen is approximately = 0" 0096.

A toisen appears to us, at the mean distance of the Moon, as an angle of 0"001046.

MAYER also calculates that the side of the Moon facing us, must has a swelling of 1/72000 of the radius, or 12.37 toisen, at the equator.

It is understood that a smaller number for the density of the Moon, than that which MAYER assumed, would produce a greater result for the swelling, which, after changing = about 167 Toisen (MÄDLER *Populären Astronomie*, 1841, p. 168.)

But that this investigation has not been exhaustively dealt with so far, we can see from an important letter of HANSENS to AIRY (*Roy. Astr. Soc. Vol. XV. 1854 Nov. 10. Nr. 1*), in which this celebrated astronomer describes his work concerning the Theory Of The Moon Report.

Though the details of them are quite unknown to me, I must not refrain from communicating an exceedingly remarkable result, though not quite certain.

HANSEN finds that the center of gravity of the Moon is 59000 meters or 8 geographical miles from the geometric center of the Moon, and the latter is closest to our Earth.

Since a Moon's atmosphere, if it exists, must be concentric about the center of gravity, or about the part of the surface of the Moon, which is nearest to the center of gravity, it follows that the Moon air would only be expected on the far side of the Moon, so that any organic life of the Moon conditioned by such atmosphere could never be perceived by us.

12. MÄDLER loc. cit. p. 13 and following, where a very circumstantial explanation of the libration explained by drawings is to be found. The effects of the Libration have been numerically discussed in detail in an interesting essay by MÄDLER: "*uber die jenseitige Halkugel des Mondes*" (Contributions to the Physical Knowledge of the Celestial bodies, Weimar, 1841, p.3) the effects of the libration have been discussed numerically.

13. WICHMANNs work on physical libration, already cited in note 6, states (*Astr. Nachr.* Nr. 631 p. 107) that:

    "The physical libration, that is, the terms of a short period, can only be very slight, so that its effect is from the Earth, less than 3", probably not even 2" etc."

    (Comp. MÄDLER Astronomie p. 170.)

14. The inclination of the lunar equator to the ecliptic found by CASSINI from his observations = 2° 30' (MÄDLER, p.11 "*der Selenographie*")

    MAYER puts it at 1° 29', and he drew this result from the repeated measurements of the Moon-mountains of Manilius, Dionysius and Censorinus.

    He calls the incliniation, α, β the selenographic latitude of the mountain, and θ the ascending node (☊) of the Moon equator in the ecliptic.

    In the 13th note of the *Kosmogr. Nachrichten* for 1748 he finds:

for Manilius	α	= 1° 29'9	θ	=−3° 45'	β	= + 14° 33'
" Dionynius		= 1° 31'5		=−0° 31'		= + 2° 55'
" Censorinus		= 1° 18'0		=+17° 30'		= + 0° 3'

From the observations of the Manilius, MAYER estimates the probable error in α = +/− 1'25, in θ = +/− 1 °.

He decides last for α = 1° 29', and concludes because of θ that probably the ☊ of the lunar equator coincides with ☋ of the lunar orbit.

The investigations of NICOLLET and BOUVARD (MÄDLER, loc. cit. p.11) yield the inclination of the lunar equator = 1° 28 '47 ", while the most precise heliometric measurements of WICHMANN are 1° 32' 9" (*Astr. Nach. Number 631 p.97*).

The former chose for their measurements the central height of the ring mountain Manilius, who chose crater Mösting A.

On the basis of the inclination found by him, = 1° 32' 9", WICHMANN says, comparing to the value from Nicollet, 1° 28 '47":

It remains undecided whether the not inconsiderable difference between the two values is due solely to the inaccuracy of the same, or is, for example, the influence of a libration of a long period.

15. A very thorough explanation of the position of the lunar equator is given by LAMBERT in the *Berliner Jahrbuch* for 1776.

16. JOANNES HEVELIUS "*Selenographia sevi lunae descriptio, etc*" *(Latin : Selenographical description of the Moon, etc.).* Danzig, 1647. In this great work, there are two full-Moon and two topographical main maps, as well as forty eight-inch drawings, representing the phase change of the Moon and the various visibilities of the mountains.

17. HEVELIUS writes in the *Selenographie* p. 151:

    "Concludo igitur, superficiem lunae clare illuminatam esse terram, maculas autem obscuriores majores esse aquas, prout quoque Kepplerus, siderali scientia clarissimus, in dissertatione sua cum Nuncio Sidereo p.29 concedit, inquiens: do maculas esse maria, do lucidas partes esse terram." —

    I conclude, therefore, that the ground surface of the Moon brightly illuminated spots are greater than the darker waters, as well Keppler, space science famous for his dissertation with the nuncio sidereo p.29 allows demon give stains on the sea give the bright parts of the Earth."

    However, it appears from other places of Selenography that HEVELIUS was not fully convinced by the oceanic condition of the grey patches, e.g. p. 149 *ibid.*

    It is not without surprise that one reads in Selenography, with what care and patience HEVELIUS (*Chapter 6*) seeks to make clear to his contemporaries why one is entitled to the assumption that the Moon has mountains and valleys and why it cannot be a mirror, of the parts of the Earth's surface.

    At that time, he had to fight not only the still important reputation of the peripatetic philosophers, but also the unmathematical objections of a CLARAMONTIUS AGUILONIUS and BETTINUS, before he came to the point with his counter-evidence.

    In particular that of CLARAMONTII opinion of Scipio, "mathematicorum omnium, philosophiae peripatelicae in quibusdam sese opponentium, jurati inimici", and that of Bettini (imprimis doctissimus Father MARIA Bettini, vehemens peripateticus).

18. LOHRMANN and MÄDLER have followed RICCIOLI since SCHRÖTER.

    MÄDLER rightly emphasizes that only the selenographer who observes himself, who has the means and the ability to use his own resources in this branch of science, is permitted to make use of the application of new names.

    If everyone had the right to attribute the name of the Moon mountains to his discretion, not only would the nomenclature lead to a terrible confusion, but vanity and flattery would be opened up, especially for those who delude them, by themselves Invention of name of selenography to render a service.

19. TOB. MAYER has not experienced the publication of his excellent map; it was later completed by LICHTENBERG.

    However, there is a very faithful copy of this in SCHRÖTER's *Selentopographische Fragmenten*, Vol. I. It has been copied many times.

    A Good one is that BENZENBERG has added to his larger works on shooting stars. GRUITHUISEN's map, which is to be based on the MAYERs, is certainly no improvement, but to the contrary.

20. J. H. SCHRÖTER, *Selenotopographische Fragmente*, etc. Vol. 1. (with 43 copper plates and 676 printed pages) 1791. - Vol. II. (With 32 copper plates and 565 printed pages.) 1802.

21. Observations by KUNOWSKY, KÖHLER, and GRUITHUISEN are often found in BODES *Berliner Jahrbüchern.*

22. W. H. LOHRMANN, *"Topographie der sichtbaren Mondoberfläche"* (Topography of the Visible Moon Surface) I. Abth. 1824. - A General map of LOHRMANN's observations, a masterpiece of the lithograph by the Patron WILHELM WERNER in Dresden, has also been published by the publisher of the present Work.
    Attempts are under way, in addition to the rich topographical material, and without harming them, by colour printing on it at the same time to represent the appearance of the Moon in full illumination.

23. The title of the great work by BEER and MÄDLER is given in note 1.
    It has 412 print pages and 5 lithographed images.
    The first quadrant of the main chart appeared in 1834 under the title:
    "Mappa Selenographica entire hemisphere of the visible Moon, from his own observations",
    and is dedicated to the King of Denmark, FREDERIK VI. Later, MÄDLER published a smaller General map and the representation of individual lunar landscapes.

24. The same artist BERKOVVSKY supplied two important small daguerreotypes of the total solar eclipse of July 28, 1851, which I had occasion to see whilst in Königsberg.
    The one picture shows in perfect clarity the corona of the totally eclipsed Sun and at the edge of the black Moon some of the red protuberances.
    During the fixation of the second image the first light of the Sun reappeared. One can see in the picture the small solar crescent, traces of the corona, and a fine fog-like hint covering the whole.

25. *Kosmos* Vol. III. P.505

26. *Astr. Nachr.* XVI p.61

27. On the occasion of the public exhibition on the Bonner "Moon relief", I have printed three small reports in some cities of the Rhine province and in Hamburg. These are:
    a. "*Das Relief der sichtbaren Halbkugel des Mondes*"(The relief of the visible hemisphere of the Moon). Olomouc 1854 Feb. 15, printed in Bonn.
    b. "*Mitteilung über ein die Gebirge des Mondes plastisch darstellendes Werk*" (Communication on a work depicting the mountains of the Moon). Olomouc 1854 July 10, printed in Hamburg.
    c. A similar report, together with a drawing in the Leipzig Illustrated Newspaper, No. 589, of October 14, 1854.

28. LAMBERT (BODES Jahrbuch, 1776 p. 151) gives only whole degrees. On p. 152 the signs of the latitudes for Tycho, Kepler, Aristarchus and Heraclides are printed incorrectly.

29. LOHRMANN's measurements on four sections of his map are to be found in the already mentioned work p.1 et seqq. Appendix.

The largest part of them, presently in the custody of the Geh. Finanzrathes OPELT in Dresden, whose splendid work, will be published later by me.

30. MÄDLERS 'selenographic observations are cite Loc. p. 69 et seqq.

There are 92 of them in total; each represents a component of 8-11 observations.

These main measurements are called "fixed points of the first order" by MÄDLER.

By connecting these (by means of triangles) to neighbouring, different points, very many "fixed points of the second order" have been determined with great accuracy.

31. *Astr. Nachr.* Nr. 631 p. 103.

32. The 27 observations, together with the individual values for Dionysius and Censorinus, are in the *Kosmogr. Nachr. und Sammlungen* for 1748.

33. According to LOHRMAM p. 93, BOUVARDS observations of Manilius in der *Connaissance des tems* (*Knowledge of the times*), 1822, published by NICOLLET in Conn. d. tems 1823.

34. 1° of the Moon equator is = 4,088 Miles,

1 '(arc) = 0.06813 miles = 259,386 toisen = 1556.30 Parisian feet.

In the center of the lunar disk, we see a degree of the equator of the Moon at the angle of 16". It is self-evident that the errors of the selenographed position determinations increase in proportion to the secants of length and width.

35. The consistency of the determinations of length and latitude at sea, when such nautical educated captains were employed, are manifested in many places in the work of Admiral KRUSENSTERN, in which he describes his memorable circumnavigation of the Earth.

A. v. HUMBOLDT's exact longitude determination of the city of Quito shows a difference of 3/4 degree from previous observations; these are observations made from Land.

Compare A. v. HUMBOLDTS *Kleinere Schriften*. Edition of 1853. Vol. I. p. 31

36. The time which flows between the opening of the upper and lower solar limb is about an hour for the equator of the Moon.

37. This is best seen in the increasing phase of the Alps, the Caucasus and the Apennines; on the Caucasus also just before the last quarter when the shadows. westwards through the Mare Serentatis.

The shadow of the easternmost mountain range of the Mare Crisium, a few days after the full Moon, is truly magnificent.

38. Excellent examples are: Copernicus, Tycho, Theophilus, and especially the sunrise over the surface of Clavius, may perhaps surpass in beauty any other scene which we can observe from the Earth on the Moon.

39. Such as Clavius, Maurolycus, Petavius, etc. These proportions from MÄDLER have long been ascertained and expressed in great and almost exhausting exactness.

40. I saw this several times in Copernicus, Theophilus, Buch, and Sacrobosco.

41. One can convince oneself of this, if from a distance the shadow of a garden wall or a gate, if such are occupied by closely spaced iron tips, viewed from above.

One will find the edge of the shadow washed out because one does not distinguish the shadows of the tips individually.

In grand scale and beauty, I saw this on the Wengernalp, when a number of ruins and towering rock fragments protrude their shadows against the dazzling snowy slopes of the Silberhorn and the large Guggigletscher.

Viewed with the naked eye, at this point the otherwise so pure shadow of the mountain ridge appears very washed out, but already with the help of an ordinary eye-glass I recognized the narrow, sharp-edged ones, and because of the sloping surfaces 20 to 50 toisen long shadows and their bright interludes clear in complete clarity.

— On the Earth a perforation of high mountain ridges, so that the Sun can shine through in many places at the same time, as far as I know, is unknown; but small-scale examples are several, For example, the South American Silver mountain Hualgayoc, (HUMBOLDT'S "*Ansichten der Natur*", vol. II, The Caxamarca Uplands); the openings (*ventanillas*) at the Guagua Pichincha near Quito (A.V. HUMBOLDTS *Kleinere Schriften*, ed., 1853, vol. I, p. 56); the Martinsloch on the Eiger, through which, twice a year, the Sun shines towards Grindelwald (G. SICHER, "*das panorama von Bern*", p.210) finally a similar gap, which I saw on one of the high rocky ridges near Altenahr in the Rhine province.

Probably, there is still a very large number.

It is just as well known that irregular or spherical very large boulders often rest freely on the crest of many mountains, whether as a result of slow weathering, or for reasons that were not explored for all cases.

42. Seen from the Earth, the penumbra has the size of 8" at the terminator.

The expansion, which is caused by a shadow of a mountain, depends on the altitude of the Sun, and is irrelevant for our measurements. (Compare SCHRÖTER *Selenotopogr., Fragm.,* Vol. 1, p. 104 et.seq..)

43. In 1831, 16 Dec BOUSSINGAULT climbed the Chimborazo without reaching the summit, the atmosphere was very clear; he had an immense view.

But from his vantage point, so close to the equator, the Sun approached the zenith at midday within about 2°, and the shadows of the mountains could not be seen.

At the time of the expedition A. v. HUMBOLDTS on this mountain (1802 June 23) the air was cloudy. (Compare A. v. HUMBOLDTS *Kleinere Schriften*, Vol. 1, p. 132 et. Seq.)

The brothers SCHLAGINTWEIT have repeatedly made observations of the mountain shadows during their alpine journey. Until now, I have not had the opportunity to get to know them.

44. I made these observations in the most favourable circumstances on the morning of July 30, 1852.

The shadow of the Eiger, which cannot develop favourably here, never appeared because the mountain was stubbornly covered in a dark blanket of cloud.

45. Through these expeditions we find short excerpts in the comprehensive book of G. Studer: "*Das Panorama von Bern*" (1850), No. 117 p. 199 et.seq.

- On p. 203-205 G. STUDER gives his own observations on the summit of the Jungfrau, whose height = 2138 toisen.

From HUGI'S [Franz Joseph Hugi] experiences on the Finsteraarhorn (height = 2193 Toisen) compared to p. 223

Considering that all these observers were surrounded by dazzling snow- and ice-fields, that during their long hours of agonizing hiking, they were always exposed to the bright reflections of the snow, it is not unlikely that upon their arrival on the summit their eyesight was in a considerably weakened or in a dulled state.

So much seems certain that the brightly lit snow surrounding them had to prove an extreme contrast to the darker terrain.

We are reminded, in this study, of the perhaps not entirely disputed argument about the colour of pure sky at great heights.

Since SAUSSURE, almost every mountain traveler speaks of the deep blue of the sky in the Alps & in the Andes.

The colour of the sky is sometimes called black-blue. However, BOUSSINGAULT (HUMBOLDTS, "*Kleinere Schriften*" Part I, p. 146) seems to be accepting that the seemingly deeper darkness of the sky is found only by comparison, that it is usually based only on the contrasts, and he remembers that according to his experiences on the volcanoes of Quito, the sky was very dark to him only when a lit snowfall was in sight.

That this assumption, though not entirely sufficient, has much truth, one also convinces oneself without having ascended the famous mountains of the Andes.

At the low altitudes of not more than a thousand toisen, the sky seemed to me so unusually dark, as I had never seen it before and after, in the direction of the deep recesses between the snow-walls of the Jungfrau and Mönchs.

The black blue color of the lakes was also the most conspicuous when the surface of the water was covered with floating, snow-covered ice fields.

On the Moon, where, of course, no snow is to be expected, the extraordinary brilliancy of the sunlit ground in the neighborhood of deep shadows will cause quite similar and perhaps even greater contrasts, and will modify the sight of the heavens or the visibility of the stars during the day.

46. In lunar eclipses, of course, the deception can occur only when the portion of the Moon, covered by the Earth's shadow, remains visible at all times, as is usually the case.

During the lunar eclipse on January 26, 1842, I first noticed the phenomenon mentioned in the text, and saw it again in similar circumstances, without paying much attention to it.

In the total eclipse of the 6th of January, 1852, which I observed at the Bonn Observatory, the apparent enlargement of the still clear bright sickle was very striking just before the Moon entered the Earth's shadow.

Comp. *Ueber das Erdenlicht* : Kosmos Vol. III. p. 497

47. *Kosmos* Vol. III. p. 499

48. Compilation of Galileo's research and discoveries etc. by Dr. R. CASPAR. Stuttgart 1854. p. 51

"*Zusammenstellung der Forschungen und Entdeckungen Galilei's auf dem Gebiete der Naturwissenschaft, als Beitrag zur Geschichte der neueren Physik*" (Compilation of Galileo's research and discoveries in the field of science, as a contribution to the history of modern physics).

49. Cosmos III. Vol. III p. 497

50. To Hamburg in May 1843

51. The first account of HERSCHEL's observations on so-called Moon volcanoes seems to be v. ZACH in Germany (BODES Jahrbuch, 1787, p. 253).

Through the reports of HORSBY[*Hornsby]* and Count Brühl, they had gained much more respect and fame than they deserved at that time.

52. The brightness of the ash-colored light is sometimes extraordinarily great, as the following examples show.

On the 28th of April, 1846, three days after the new Moon, I watched on the night a slight enlargement using the eight-foot Bonnner heliometer.

If I removed the crescent Moon out of the field of vision, I saw the remaining part of the sphere in a very bright gray-yellow light, in which the individual points of light belong just as little as in the full Moon.

On the whole, the sight was quite consistent with that given by the full Moon, when the latter is covered so far by its cirrus or snow-clouds that the weakest shadings of the surface are already lost from sight.

On February 18, 1847, I watched the two-day-old Moon at the five-footer Refractor of the Bonn Observatory.

The *lumen secundarium* was so splendid that, when I covered the narrow cresent through the ring micrometer, I clearly recognized the whole circumference of the Mare Crisium, although the western boundary-range of this plain was already illuminated by the Sun.

— At the points where, seen from the observatory to Bonn, the Moon can rise and set, there are mountains of different heights on the horizon.

One never observed the Moon in zenith distances of 90 °.

But I noticed several times that the ash-colored light remained visible to the naked eye, when the crescent Moon, which was 1° to 2° high, was covered by mountains.

In Sept. 1852 I observed several times the rising of the waning Moon over the horizon of the North Sea, on the Frisian island of Foehr.

I noticed the ash-colored light earlier than the cresent, but I add that possibly the Moon rose for my point of view over the extremely flat archipelago, a few arc minutes over the apparently high Jutland coast, and not far above the sea horizon.

In the course of the day, the coast was frequently extended beyond the horizon by the action of refraction, sometimes separated by a strip of air; but the darkness of the night prevented us from deciding whether the coast was below or above the horizon at the time of the moonrise.

The height of the eye above the sea surface alternated between 1 and 6 toisen.

If we consider the most significant extinction of light in the atmosphere, which makes fixed stars of the first magnitude invisible directly on the horizon itself, then at certain times we must ascribe a very great brightness to the *Lumen secundarium* of the Moon; of course, the large area contributes greatly to our being so close to the horizon.

Incidentally, such observation is one of the rarest in our Zone.

If one rightly attributes to the continents of the Earth a much greater brightness (seen from the Moon), and derives the variable intensity of the ash-colored light from the change of the light-reflection from these and from the oceans, then it is not to be overlooked that during winter months the northern hemisphere of the Earth is even more suitable because of the snow cover.

Let us suppose that the temporary winter boundary of the snow extends only to a latitude of 55° N. From the Moon, the snowy land zone of northern Europe, Siberia, and North America is already considerable, and as in the winter of + 65° of latitude the sea, in combination with the above-mentioned land-zone, offers a very large surface of intense white which surrounds the pole, and which appears from the Moon in not too great a contraction.

This temporary snow zone will very often be covered by clouds; however, the sunlight reflected upwards or downwards by the cloud is intense enough to throw more light upon the Moon than a continental surface can do for itself.

As far as the South Pole of the Earth is concerned, it is not fully known how far the fixed ice cover extends to the north at the time of its winter.

The stable ice wall discovered by JAMES CLARKE ROSS, which towers high above the sea to the South Pole, was 781/2° wide at the time of summer.

53. BEER and MÄDLER, "*Beiträge zur physischen Kenntnis der Himmelskörper*" (contributions to the physical knowledge of the heavenly bodies). Weimar, 1841. p. 43

The magnification coefficient was calculated from the spot observations in 1833 and 1837.

By contrast, in June 1835 MÄDLER measured the shadow of the very small lunar eclipse micrometrically, and deduced the result, which deviates quite a bit from all others = 1/28.

It seems to me that the micrometric determination of the penumbra will always be of value, which will be consistent with those determined from spot observations.

This seemed to me so on the occasion of the partial lunar eclipse of 6 Dec. In 1843 (Hamburg), and even more so in the still smaller one of 4 November 1854 (Olomouc), when I remained utterly doubtful, and indeed to 1 and 2 arcminutes, the size of the micrometer.

As long as our knowledge of the enlargement of the shadow of the Earth is still very limited, it would be good not to put too much emphasis on the value of 1/28. — In LITTROWs 'report on the MÄDLER's Map (J. J. v. LITTROW's V*ermischte schriften*, edited by E. L. v. LITTROW, Vol. II, p. 142), MÄDLER is said to have found that magnification = 1/28, without mentioning the other two more precise values of 1/65 and 1/54.

Comp. in the yearbook, of Lit. 1838. LXXXII.

The last of MÄDLER's indications in the text = 1/49, can be found in the *Astr. Nachrichten* No 527.

54. My own observations and calculations, which have not yet been completed, which are to be attributed to the magnification coefficients, can only be fully communicated in a later period.

I give the results here more specifically. If, as before, I call the questionable value A, I have found:

1. Partial Lunar eclipse on 26 January 1842:

A = 1/48 from	18 observations from me from Eutin in Holstein,		
A = 1/49 "	16 "	"	MÄDLER in Dorpat,
A = 1/54 "	8 „	"	RÜMKER in Hamburg.

Average with respect to the number of observations = 1/50, valid for the ADAMS improved parallax = 61'24"6.

2. Total Lunar Eclipse on 31 May 1844

A = 1/53 from		17 Observations from me in Hamburg,	
A = 1/61	"	11 "	RÜMKER in Hamburg,
A = 1/49	"	7 "	FUNK in Hamburg,
A = 1/57	"	13 "	GAUSS in Göttingen,
A = 1/57	"	9 "	GOLDSCHMIDT in Göttingen,
A = 1/60	"	7 "	QUETELET in Brüssel,
A = 1/63	"	9 "	LIAGRE in Brüssel,
A = 1/61	"	3 "	GERLING in Marburg.

Average = 1/56 for the parallax = 61' 13"3.

3. Total Lunar Eclipse on 24, Nov. 1844:

A = 1/49 from	37 observations from MÄDLER in Dorpat (Astr. Nachr. Nr. 527)		
A = 1/61 "	6	"	DE VICO in Rome

Average := 1/52 if I add MÄDLER's result to 3 times the value.
Parallax = 53'59"9.

4. Total Lunar Eclipse on 19 March 1848

   A = 1/45 after 17 observations from me in Bonn. Parallax = 54'23"0.

5. Partial Lunar Eclipse on 8. March 1849:

   A = 1/44 after 15 observations from me in Bonn. Parallax = 56'51"7.

55. The reports about the colour of the Moon during an eclipse are quite numerous.

Even KEPLER and HEVELIUS (*Selenographie*, p. 116) generally knew the cause of that colour; HEVELIUS says:

"..... constat, lumen illud debilius vel colores in eclipsibus apparenies non ex proprio lunaeque insüo lumine oriri, sed a solis luce, una cum lumine secundario lunae, quod circa novilunium animadverlitur .... "

< This is evident, that faint or colours are seen in the light of the partial eclipses are present in the light of the Moon 'visions appearing to them to arise, not from man's own, but only from the light of the Sun, the Moon, together with the light, in the second place, that it is turned with regard to the new Moon, to give attention to>

Compare. *Kosmos* Vol. III. p. 500.

On the disappearance of the Moon in total eclipses, HEVELIUS p. 117 et seq. (*Kosmos* Vol. III. p. 499.)

— Similar things were seen in this century 1816 and 1823.

Although the data on the Earth's shadow may be numerous, they are generally too general, and often accompanied by remarks of a peculiar kind.

Thus HELFENZRIEDER reported that in the darkness of July 30, 1776, he had seen the shadow of an unequal curvature, and in it bright streaks, especially at Tycho.

This sounds like HELFENZRIEDER considered the stripes to be a special feature of the Shadow, and did not know that this very distinctive phenomenon is unique to Tycho.

If he knew the Latin, the HELFENZRIEDER observation, (which can be found in the German translation in BODES *Jahrbuch* 1779 p. 43) might be different.

The large unevenness of the shadow boundary, repeatedly cited by the observers, was certainly based on an illusion (because of the alternation of bright and dark spots on the Moon), that SCHRÖTER (BODES *Jahrbuch* 1794 p.120) on occasion of the total eclipse of October 22, 1790 has beautifully explained.

It can be seen from other Reports (before p.108) that the blue light was already known at that time.

-On May 31, 1844, the totally eclipsed Moon seemed to have a greenish colour beside the blue light.

By the way, it should be noted here that a formerly well-known astronomer was quite naively quoting: "he could see stars of seventh magnitude through the shadow of the Earth!"

56. This was, for example, 1848 March 19 and 1852 Jan. 6

In passing mists, a reddish corona of small extent is formed.

In order to obtain the halo of 22° radius, however, the eclipsed Moon was much too bright, although the crystals of ice, which caused the halo, would have been present in the air.

57. The number of observations on total solar eclipses is now very large.
I limit myself to a few examples.

1778 June 24. Admiral ULLOA Observations on the sea. BODES *Jahrbuch* for 1781. Also. 1782 p.144, 1794 p.106

1842 July 7. Reports in the *Astr. Nachrichten* of that time, and in the annals of the Vienna Observatory.

1850 August 8. KUTCZYCKYS Observations in Honolulu. (*Compl. Rend.* XXXIII No. 16. April 21, 1851.)

1851, July 28. The very numerous reports of these total lunar events observed in so many places are scattered in astronomical news and in English astronomical journals

The number of monographs is not inconsiderable. The English observations have been collected in 1853 in a special volume with many illustrations published.

The observations of the French astronomers MAUVAIS and GOUJON are also found in the Astron. Nachr. and other printings.

The monograph of my own observations on Rastenburg, provided with four illustrations, was published in 1852 at the instigation of the royal court Observatory of Bonn.

1852 Dec. 10. I have no more information about the total eclipse in China.

1853 Nov. 30. Dr. MOESTAS Observations on Pisco in Peru. *Astronomical Journal* No. 67. Cambridge U. S. 1854 Feb. 17. "Letter from Dr. MOESTA, Director of the observatory at Santiago de Chile to Lieutenant GILLIS". In JAHNS weekly astronomer.

"Unterhaltungen" Conversations (Autumn 1854) we find the German translation of the very estimable Peruvian observation, the only one which has become known to me so far.

The main report on this eclipse is: Eclipse solar de 30. de Noviembre de 1853. Presentado al Seuor Ministro de Instruccion publica por CARLOS MOESTA. Santiago de Chile. Marzo de 1854 (with a lithography table). *[Report on the observations made during the Solar Eclipse of November 30, 1853. Presented to the Minister of Public Instruction by CARLOS MOESTA. Santiago, Chile. March 1854]* - compared nor is it ARAGO's essay in the Memoire du Bureau des Longitudes, 1846, and Cosmos, Vol. III. p. 378 et seqq and p. 488

58. Astr. Nachr. XI. p. 411 et seq. MÄDLER loc. cit. p. 133. p. 152. For the sake of curiosity, allow TOB. MAYER's opinion about the lunar air.

BESSEL wrote down his view in 1834, MAYER his, 86 years earlier. *Kosmogr. Nachr. und Samml.* IX. p. 397 ("Proof that the Moon has no air,")

MAYER wrote:

"I will be accused, as if I had in mind, with the air of the Moon at the same time to destroy its inhabitants. And in truth, I confess that I have a great desire to do this, if I knew that this would give some benefit to the confirmation of my sentence.

Alone, I see a means of ridding myself of this cruelty, and of those who are disposed to live and rational creatures on the Moon, of a grief. To be alive and to be reasonable, in my opinion, does not so much conflict with the lack of respite, of speech and of hearing.

Of course, the last things must be missing from the lunar inhabitants as soon as they take their breath away. That's the only reason they can survive.

For my sake you can have as much and more reason than we do; the language may have been replaced in another way; their bodies may have been made by the Creator, that they are able to endure in an infinitely thinner matter than our air is. Just as we see in the fish that live in the water, in a matter that is a thousand times denser, just as healthy and comfortable as we are in the air. In short, I give everyone the freedom to put creatures on the Moon and to judge what kind of physical condition they may be, only I ask them to make them in such a way that they need no air. "

— reference also the earlier note. 11, where HANSEN's investigation of the position of the center of gravity in the Moon is discussed.

59. The expression "formation" in the sense of the geologists may not be applied to the lunar mountains without further ado, because we know nothing of formations of the mountains there, we will hardly ever know anything else.

Only observers who tell the public the brilliant result of having seen on the Moon layers of 30 feet in thickness, while their telescope was not yet able to visualize dimensions of 300 toisen, may apply the term in earnest.

One finds about this geological expression in Karl Cäsar von LEONHARDs

"*Lehrbuch der Geognosie und Geologie*" (Stuttgart 1846-49) p. 220 et seq. The following explanation:

"With the word formation, one connects, according to A. v. HUMBOLDT [*Geognostic test on the rock deposit, page 1]* has a double meaning. Once this is understood as the type of formation of a rock; then a system of mineral masses, which in such a way are connected with each other, that they may be considered as being created at the same time, that they, even in the most remote regions, the same general conditions, which indicates storage and disposal. Thus the formation of the obsidian and of the basalt is attributed to the effect of underground fire; it is said that the formation of clay slate surrounds pebbles and slate, and layers of black limestone.

The first meaning of the word is more in keeping with the spirit of the language; but it refers to the origin of things, to an uncertain knowledge based on geogonic hypotheses, on sets of propositions, conceiving the Earth-formation-doctrine.

- The second meaning of that expression is borrowed from the school of WERNERS;

he confines himself to what is and does not deal with requirements as it might have been. "

In the consideration of the lunar mountains one must only speak of formations or structures and of directions. Also the mining term "strike and dip" is to be avoided.

Surface design, height and horizontal dimensions, mutual position, and with many craters the relative age alone are objects of an ingenious research, and only from this one is hoped that it will significantly promote our knowledge of the lunar mountains.

60. MÄDLER loc. cit. p. 99.

61. MÄDLER loc. cit. p. 90. and SCHRÖTER, "*Selenotopographische Fragmente*" Vol. I., p. 74 et seq.

62. A. v. HUMBOLDT, "*Voyage auxiens equinoxiales*" (*Travel to Equinoctial Regions)* T. 1. p. 276. and DEVILLE's Journey p. 6 (treatise on Tenerife).

63. A. v. HUMBOLDT, "*Kleinere Schriften*" v. 1853 vol. 1. p. 165, notes that for BORDA's trigonomic measurement of the Pic de Teide, the height was erroneously calculated in 1771, because one of the angles, instead of 53', was entered as 33'.

64. This barometer measurement BORDA's has v. HUMBOLDT calculates according to the LAPLACE formula.

65. These can be found in HUMBOLDT's "*Ansichten das Natur*" (1849) Vol.2 p.290

66. Neither A. v. HUMBOLDT (23 June 1802) nor BOUSSINGAULT (16 Dec 1831) reached the summit. The expeditions to the Chimborazo are described in the "*Kleinere Schriften*" Vol. I. The three heights of the mountain are found in the same work p. 167

67. Among seven different trigonometric combinations, the difference between the largest and smallest height was only about 2 toisen.

Compare *Kleiner Schriften* Vol. I. p. 163 et seq. Ditto p. 159 speaks of v. HUMBOLDT about the measurements of Montblanc and Monte Rosa. where the deviations are resp. 1/246 and 1/34.

In the great work of Herman SCHLAGINTWEIT. and Adolf SCHLAGINTWEIT "*Neue untersuchungen über die physikalische Geographie und die Geologie der Alpen*" (New Investigations on Physical Geography and the Geology of the Alps((1854), one finds a very interesting summary of the exact height-determination of Monte Rosa and many other mountains.

68. SCHRÖTER's measurements were made at the end of the last century; MÄDLERS between 1830 and 1837; my measurements since 1844. The mountain of Calippus had only once been given by SCHRÖTER, namely 1788 Nov. 6 (vol. I. p. 266).

69. I limit myself here to my observations, because it seems that neither SCHRÖTER nor MÄDLER could trace the fine shadow tip to the end. Moreover, with MÄDLER (p. 123, No. 1085) φ is certainly distorted by a pressure error; even at 3°12' sunshine, the shadow tip is not yet separated from the phase.

70. Probably it will never be possible to determine the maximum height of Curtius δ over the main crater through the shadows because, as MÄDLER loc. cit.. p. 402 supposes that the mountain is somewhat rounded, though not much, in the form of a dome.

In twelve Lunations I have tried in vain to observe the separation of the shadow from the phase during a waning Moon; ten times it was cloudy and twice the libration did not allow a clear measurement.

The shadow ends south-west of Clavius in the wild mountain range. I found in an observation that it is not yet separated from the phase at a Sun altitude of 5° 29', from which it follows that it is a height of more than 4000 Toisen over the external region.

As the bottom of the Curtius crater lies deeper than the whole environment, the summit above the center of the basin must be more than 4000t; but when the shadow has reached the center, the Sun is still 10° to 11°, and it is to be assumed that the uppermost parts of the mountain will not cast a shadow.

However, there is a prospect of ascertaining the greatest relative height another way, and if this is (calculated eastwards) as much as 4,000 toisen, it would not be surprising if the inner relative height = 4800t or even 5000t.

71. On the great map of MÄDLER this central mountain is missing; however, in the text of *Selenographie* p. 398 expressly remarked that he was only omitted by chance. LOHRMANN has also listed it.

72. "*Views of Nature*", Vol. II, p. 256 and Cosmos, Vol. I, p. 218. The Yorullo was founded on 29 September, 1759, and Monte Nuovo on September 19, 1538. –
Cosmos, Vol I. p.251. The collapse of the Carguairazo on the 19th of June, 1698 (*Kosmos*, Vol. I, p. 243, and *Kleiner Schriften* Vol. I. p. 137 on the Capac-Urcu.)

73. These large snowy mountains are pictured in the beautiful atlas by A. v. HUMBOLDT's - *Kleiner Schriften.*

74. BERGHAUS *Geographie Jahrbuch* 1851 Book III. p. 21 and HUMBOLDT *Kl. Schriften* Vol. I. p. 163, where the height measurements in the Himalaya are discussed.

The greatest sea depths discovered by captain DENHAM 1852 Oct. 30 in the southern Atlantic Ocean, 37° 6' west longitude of Greenwich, and 36° 49' south latitude (HUMBOLDT in the *monthly report of the Royal Prussian Academy of Sciences*, Feb. 1853, p.140).

7707 fathoms (7.7 miles/44,000 feet) on 20th October. Took nearly 9.5 hours for the 9lbs plummet to reach the seabed.

75. The expression "passage-like" must not be confused with the same geognostic expression.

A passage is a more or less important cleft in the rock of some mountain which is filled with foreign rocks.

For example the chalk and the limestone of a sedimentary structure, a precipitation from the water.

Later, when the sediment had solidified, it split by metamorphic uplifts, and the new fissures were filled in by the ascending, fiery liquid (metamorphic) material;

we see passages of basalt, porphyry, and granite as fillings in the normal rock, and sometimes they have overflowed at the outgoing end, and, like the head of a mushroom, have spread round the surface of the mountain.

When the word "passage-like" is applied to the Moon, one must first think of the matter, which is lies horizontally beneath the surface, and which, as long as it has resisted, only buoyed and loosened the surface, and probably also metamorphosizing the structure and the colour.

If one takes this passage-like direction of the force as a tubular erosion, then the effect of this is very similar to mole tunnels.

76. With a completely convincing certainty, I cannot prove a totally wallless pit on the Moon.

77. There is no possibility that a gorge, like the natural rocky bridge of Iconozo, might be overhung by a precipice of rock.

This may happen often enough; but it should only be emphasized that we know nothing about it, that no one has ever seen anything like this, and (for obvious reasons) never will something like that happen.

78. As is well known, in the mountains of our Earth, the "shifts, bends, curvatures and fractures" of the strata are very frequent, both small and large. But we must distinguish whether an isolated mountain has been brought out of its position by a force which is close to it, or whether a certain place, a certain Mass of a whole mountain range, has been warped or twisted. The first case, as a separate phenomenon, cannot be demonstrated, provided one abstracts from the fact that in a whole mountain system, as in the Alps, the aging of falling and striking points to such occurrences in prehistoric times; but the examples for the second case are numerous. Thus, a local, more or less great deflection of individual layers of rock in a whole mass of mountains, such as the admirable example at the mouth of the little Lütschine in the alpine valley of Lauterbrunnen, cannot be taken into consideration for the Moon, but a lateral displacement, or even an overhang of crater walls, or isolated elevations, and these already by observation many times indicated conjecture will one day probably be raised to certainty.

79. One is reminded involuntarily of the bumpy Malpays, which was first climbed in 1759, in the midst of its back, the larger masses of the Yorullo, and the small round parasitic Hornitos.

The small craters, which stretched from north to south over the expanded surface of Mersenius, remained unknown to the former observers; there are 7 or 9, and in between the hills. Only once, on 25 April 1797, some of these fine objects were seen by SCHRÖTER and HARDING. (See, for example, *Selenotopogr. Fragmente*, Vol., § 664, p.118)

80. Coloured areas in which dark and light gray alternate, we find in Schicard, Mare Humboldtianum in the Alphonsus and in Petavius.

81. This mountain is probably η according to MÄDLER, but the letter is missing in the *Mappa Selenographica.*

82. The Berlin observation took place in May 1853.

83. First seen using the Berlin Refractor in May 1853.

84. In favourable circumstances, I counted, in the area of the Clavius, 90 craters (May 17, 1853), including the largest ones.

85. MÄDLER has not measured this peak. Like everything in this area, it is difficult to determine because of the irregularity of the phase. I found:

with	H = 4° 28'	h = 2200t	from 3 observations	
"	H = 5 29	h = 2621"	4 "	
"	H = 7 26	h = 2887"	2 "	
"	H = 9 09	h = 2658"	4 "	
"	H = 9 56	h = 2490"	4 "	
"	H = 13 15	h = 1922"	2 "	

86. According to the measurements of MÄDLER; the results are certainly too few to contribute to the decision.

87. I have determined the western side of the Copernicus as follows:

S -West-wall	with	H = 5° 12'	h = 1749t	from 3 Observations
W.-Pic on the wall		H = 6 08	h = 2047	" 4
" " "		H = 9 53	h = 1928	" 3
" " "		H = 10 40	h = 1641	" 1
" " "		H = 15 45	h = 1082	" 1
W.-wall, close north from Pic		H = 7 02	h = 1902	" 3
N.-West-wall .		H = 5 38	h = 1852	" 3

88. Theophilus is often difficult to measure due to the phase of the Moon. I found:

S.-West-wall with	H = 7° 51'	h = 1733t	from 3 Observations,
West-wall "	H = 5 31	h = 1915	2
" " "	H = 7 55	h = 1985	1
" " "	H = 8 03	h = 2025	2
" " "	H = 8 25	h = 1836	2
" " "	H = 9 37	h = 1553	2
" " "	H = 12 31	h = 2245	1
N.-West-wall	H = 5 08	h = 1790	1
" " "	H = 7 22	h = 2250	1
" " "	H = 8 03	h = 1987	2
N.-East-wall	H = 7 14	h = 1547	2
S.-East-wall	H = 7 27	h = 1993	2
East-wall with	H = 7° 34'	h = 2062t	from 2 Observations,
"	H = 1 54	h = 2328	2
"	H = 10 05	h = 2696	1
"	H = 13 30	h = 2212	2

89. The location of these craters is according to the LOHRMANN's co-ordinate system.

The greatest one, southwest of the wall of Hesiod, has 17° East longitude and 29° 7' South latitude. LOHRMANN uses the number 61.

The smaller in the Sinus Epidemiarum, 29 ° east longitude and 31 ° south latitude, LOHRMANN provides no designation.

The former is, according to MÄDLER's Map = Hesiod A, the second is unmarked.

90. The *Mappa Selenographica* contains approximately 7800 Ring mountains of all sizes and shapes. If I count only the larger ones, about two and one and a half miles downwards, I obtain, according to a very casual estimate, in which all the small craters were not counted:

In the	I. Quadrant	c300 Ring mountains,	
" "	II. "	230 "	"
" "	III. "	850 "	"
" "	IV. "	1200 "	"
Total		2580.	

So there are still 5220 smaller ring mountains left over. If I now examine how many craters I have seen more than MÄDLER, with the medium optical power of my telescope, for Clavius 3.7, for Ptolemy 2.7, for the area near Stadius 2.2 times more craters, than is contained in the *Mappa Selenographica.*

It would therefore be concluded that even in five-footed telescopes with a magnification of 200 to 300 times, after a long observation under favourable circumstances, 19,000, 14,000, and 12,000 ring mountains of all sizes would be recognizable on the Moon.

However little determined and so crude this estimate may be, it is nevertheless to be understood that for a 14-foot refractor probably 50,000 craters should be visible.

91. I attribute this small height to the smallest craters that can still be recognized; the larger ones sometimes permit a direct measurement of their outer wall height e.g.

Galilai	= 341t	Sulp. gallus . .	= 204
Galilai a . . .	= 247	Crater bei Arago .	= 183
Delisle c . . .	= 322	Marius C . . .	= 127
Delisle 6 . . .	= 257	Crater bei Pico A	= 157
Crater Pico A . .	= 342	Heraclid e . . .	= 119
Crater Pico B . .	= 283	Bessel d	= 85
Sosigenes a . . .	= 248	Seleucus A . . .	= 79
Timocharis d . .	= 266	Briggs A . . .	= 66
Taquet [*Tacquet*]	= 262 t		

The last three, therefore, reach only the height of our largest church towers, and very many small craters certainly do not have half of this height. Their depth cannot be determined.

92. A small crater on the mountain of Huygens is 2800t higher than the Mare Imbrium, but the profile of this long, miles-long, steep mountain range is not comparable with that of our isolated conical volcanoes.

93. From the Great opus of LOHRMANN's, I shall examine a few cases of this kind in more detail.

94. In the contributions to the physical knowledge of the heavenly bodies (Weimar, 1841), MÄDLER has included the 92 rille known to him into a catalog and described them.

95. If I attribute to SCHRÖTER the knowledge of five rille (which MÄDLER does not do), I calculate here, for the time being, the great cross-section of the Alps with the rille;

the curved furrow at Aristarchus and the rille east of the straight mountain wall at Thebit in the Mare Nubium was already observed by SCHRÖTER. (SCHRÖTER, loc. cit. Volume 2. P.342, § 908. 2.)

96. This number may be greater or smaller, since at the time when I made the count, several plates of LOHRMANN's maps had not yet been completed, and the original was not yet carefully compared.

97. KINAUS Moon observations can be found in JAHN's *Wöchentlichen Unterhaltungen* in 1848. No. 25. p. 202 et seq.

98. The rille of Rhaeticus and Réaumur were first seen by the Berlin Refractor in May, 1853.

99. I also recognized the grooves of Arzachel and Alphonsus only using the Berlin telescope. LOHRMANN already noticed a trace of the groove in Arzachel.

100. From the Berlin observations in May 1853.

101. From the observations made in Olmütz [*Olomouc*], 1854.

102. A measurement attempted on the 7th of July, 1854, revealed the depth of the great rille of Aristarchus, north at E = 260 toisen. But in most places it has a much lesser depth.

103. This is not completely accurate. Some of the rocks of Pichincha and Chimborazo can be found in various places in v. HUMBOLDT 's *Kleiner Schriften* of 1853, vol. I, e.g. p.159

104. In the abovementioned Alpine panorama of G. STUDER, one finds the type of rock given for every main peak, and it shows how differently the composition of neighbouring mountains may be.

E.g. says STUDER p. 206: "Geologically interesting is the fact that during the same high mountain limestone, which serves the base of the Jungfrau, the whole mass of the powerful Eiger, the southern Mönch remains in the area of the Gneiss region."

105. According to an estimate by MÄDLER, cited . p.241.

106. MÄDLER has already drawn attention to these and many other strange relations of coincident-layers and other layers 20 years ago.

107. Length of the eastern border of the Montes Alpes = 34 miles according to MÄDLER.

108. Caucasus, about 37 miles long, and according to MÄDLER about 450 square miles of area.

109. This measurement, to which A. v. HUMBOLDT has often referenced, has been attempted for some mountains of the Earth.

One wants to express the measure of the lifting force by the ratio of the middle ridge height to the highest peak height.

In the Atlas of A. v. HUMBOLDT's *Kleiner Schriften*, contains a description of these ratios, which, when the mean crest height is set = 1, yield the following numbers:

	Average crest height.		Max peak height.
Alps . . .	= 1	:	2.05
Pyrenees	= 1	:	1.43
Andes . . .	= 1	:	2.02
Himalaya	= 1	:	1.79

The mountains with the largest peaks in these four ranges: Montblanc = 2468t, Pic de Néthou = 1787t, Aconcagua = 3722t, Kintschinjinga = 4406t; as mean ridge heights in the same order the numbers: 1200t, 1250t, 1850t, 2450t.

110. The very instructive essay of HODGSON on the Himalaya (German translation) can be found in BERGHAUS *geogr. Jahrbuch* vol. III. p. 20 et seq., and the small map (No. VIII) in the same book.

The great map of the Himalaya, according to WEBB, HODGSON, and HERBERT, edited by BERGHAUS, is given in book II of the same work of 1850; but with the exception of the southern (Indian) mountain drop, only the individual peaks, but not the general configuration of the mountains, are found here.

It terminates at 81° 20' east longitude from Greenwich, and thus no longer contains the Eastern part of the Himalaya, which HODGSON represents on his small map; p. 29 HODGSON gives a profile of the Southern slope of the Himalaya chain to explain three different regions.

These are:

1. *The Terai*, the swampland, slightly submerged under the Indian plain, at the southern foot of the whole mountain range.
2. *Bihar*, first ascent of the mountain range from the Tarai towards the north, a woody debris and ruined land.
3. *Dooars*, sandstone region, the higher level above the *Bihar*, a flat trough-shaped valley, separated from the *Bihar* by a little high sandstone elevation separation.

From this last region the mountains rise more and more to the central mass, to the Ghat or snow line of the main ridge. But this ridge does not descend to the plain in the same way (north) as the high peaks mentioned in the text; there is a land of 1400t to 1800t, and the opinion has been expressed (but perhaps not yet confirmed) that an even higher mountain range would rise further north. The investigations of the Himalaya are only in their beginnings. More recently the brothers A. and H. SCHLAGINTWEIT have already begun their researches there.

111. One does not have to believe that only the peaks mentioned above are available. The small special maps of the *Sikkim-Himalaya* by Sir JOSEPH DALTON HOOKER (in the works of BERGHAUS, cited under note 110) shows that in the vicinity of the highest snowy mountain Kintschinjinga there are several other mighty peaks; so, for example,

3.7 miles west of Mount Nango = 3284t;
1.1 miles west of Mount Junnoo = 3958t;
0.2 miles southeast of the main summit, a secondary dome = 4351t;
1.6 miles southeast of Pundeem = 3442t;
in the same direction 2.7 miles away the mountain Nursing = 2993t and so on.

So here mountains of 2000 to 3000 Toisen high and over are frequent. For the Himälaya HODGSON takes a length of 450 and a width of 22 geogr. Miles.

Nanda-Devi is identical to the Jawahir. - Comp. also on the Himalaya: HUMBOLDT in the treatise on the mean height of the Continents, *Kl. Schriften* Vol. I. p. 425 et. Seq.

112. I do not want here to have called the Niesen an isolated cone;

it appears, indeed, from Bern, and notably on the surface of Lake Thun, on whose south-western shore it rises, but as the most northerly member of the Niesen chain, the beginning of which is to be sought on the Strobel.

113. To this remarkable, fourteen-mile-long, curving wall one hardly finds a second counterpart on the Moon as on the Earth. MÄDLER finds the height = 157t, I from 2 measurements = 138t.

114. It depends very much on the state of our atmosphere, and even more on the experience of telescopic vision, whether we recognize the weak green color in the Moon plains or not. I noticed them first (without knowing anything from MÄDLERs much earlier detection) in May 1843 at the Hamburg Observatory, in Mare Serentatis.

The air was by no means favourable, for it was, if cloudless, lightly covered with haze (high altitude haze). It did not appear to me in some observations that the green color had lost or changed immediately after the end of lunar eclipses. A brownish color is often seen clearly in the hill terrain northeast of the Aristarchus in the area of the multiple-curved rille; I have never been able to perceive the reddish shimmer at Lichtenberg, of which MÄDLER speaks.

115. I first saw this strange crater during an observation of the Moon at the Meridian Circle of Bonn at the beginning of 1851; I noticed for the first time its gray triangle. There are a few other craters who show something similar. But whether the phenomenon is changeable cannot yet be determined.

116. Comp. MÄDLER's *Selenographie* p. 136.

117. Many large elevated craters of the Earth show on their slopes numerous barancos, which, like the cracks of a window pane, pierced in the middle, spread from every point in all directions. L. v. BUCH therefore called this phenomenon an étoilement. The Barancos begin outside close to the crater walls, and, in width and often in depth, they move on all sides against the base of the mountain or against the sea. The uplifted, originally horizontally supported mass had to split many times during their erection in the direction indicated by the Barancos. Some cracks closed again; most remained open, and received their present form by the atmospheric water.

Whether or not the fissure of the outer crater wall of the newly-formed mountain may proceed in the manner mentioned in the case of the raising of a large, homogeneous mass of rock has not yet been discussed either by great experiments or by a strict scientific investigation.

Sometimes a baranco breaks through the crater wall; it is on the S. W. side of the mighty craters on the Canary Island of Palma, where one can descend from Caldeira through the great main valley (Baranco de Augustia) towards the port of Tazacorte. The illustrations of the islands of Palma and Tenerife in L. v. BUCH's great work on the Canary Islands, are very beautiful and clear, and also the seacharts of the Azores and the Canary Islands, published by the Hydrographic Office in London. The crater of Palma shows at least forty such valleys; but the number of them is still greater in Tenerife, where they descend from the edge of the presumably old craters (from whose surface now the Pic de Teide rises), especially to the south. —The Azores islands also give excellent examples of this type, such as the Caldeira St. Barbara on Terceira, Fayal, Corvo and the great Caldeira das Sete Cidades at the west end of San Miguel.

According to DEVILLE's Map, the numerous Barancos of the Cape Verde island of Fogo, which, like Palma, appear only from a single large mountain, the elevation crater, to the old ground of which the still extinct volcanic cone of Fogo rises, are very pronounced; on the northern slope of Mount Vesuvius, or more precisely, on the outer surface of the Somma, the numerous valleys descending to the plain may at least be compared with the Barancos, and (according to the ABICH's map) many such ravines on the extreme slopes of Roccamonfina, north west of Naples. The views of the geologists on this subject are, however, too great, and I am very far from wanting to express an independent opinion here. It may be recalled that, as regards to the crater, Roccamonfina, Sir RODERICH MURCHISON considers that the mountain is of submarine origin, and that a deposit cone is formed from slag. In this case, the explanation of the Barancos would be removed if the opinion of MURCHISON is the correct one. In general, it is difficult to find the truth in the often so complicated circumstances; the old Roderberg, a volcano near Bonn, shows, on its northern slag, which is amenable to the Rhine, 7 to 9 wavy folds, which have sometimes been interpreted as a lava flow from a superficial view. Of course, this is unthinkable. But from a distance, e.g. Seen from the heights of the opposite Siebengebirge, the Roderberg is the only one in the whole area, in which faint valleys descending from the height towards the foot have a distant resemblance to the Barancos. But this resemblance is only feigned.

In Palma, the Barancos form deep and rugged ravines in the metamorphic rocks, while for the mentioned valleys on the Roderberg, where below the dividing wall of each two furrows is intersected, the explanation is to be found. The land, which has been flooded and deposited in the mountains, is situated in front of the Loess formation, which, through unknown processes, received the undulating fold of its surface.

What the composition of the walls of the Roderberg is underneath is unknown. Comp. BERGHAUS *Geogr.. Jahrbuch* 1852 IV. p.62 and THOMAS dissertation on the Roderberg.

118. The view, however, is not the same at different times. The finer observation made in very favourable circumstances shows Aristarchus, surrounded by a broad, angular nimbus, in which the radiation of the stripes can already be seen. They enter the crater and begin at the Central mountains. As they rise on the crater walls, they appear separated by grey spaces, and thus this brilliant crater is as colorful as some of the other kinds.

119. Comp. with MÄDLER loc. cit. p.195.

120. T*he* average (middle) of the Caldeira of Palma is here approximately from N.W by S.E. placed by the peaks Pico de los Muchachos and Pico del Cedro

121. The direction of the average goes through Garachico (north) and Abona Point (south).

However, this is not all too accurate, and even less the fact that I have even included the Chahorra. It was only important to make clear the conditions of the mountain approximately.

122. These are arbitrarily chosen examples, to which no particular objects correspond in reality. The average coincides coincidentally with the upper part of the volcanoes of Manderscheid in the Eifel; but just these are very different from the Maaren of the local area in every respect.

I am not in a position to judge a profile which is based on measurements.

The drawings, which I brought with me from the crater-rich Eifel in 1847, depict only landscape, not geognostic, several of the large maaren.

123. According to DEVILLE's paper.
Études géologiques sur les îles de Ténériffe et de Fogo (1848)

124. A walled area, for instance that of Ptolemy.

125. A big crater for example Copernicus or Theophilus.

126. No.1 e.g. the crater Geber No. III of crater Hesiod *A*

127. I note that these figures are only taken from the Map and the profile of Fogo (after DEVILLE), and quite incidentally, like many of the rest, are not to be taken too seriously.

I have chosen the western corner from the Circumvallation; but as far as the depth is concerned, this is of no value in the present example, because the actual volcano, the Pico de Fogo, fills the whole of the crater, so that one can probably quote the height of the old walls above the base of the pics, but not over the former depth of the protruding crater. The same applies to Mount Vesuvius, to the Pic of Tenerife and to Roccamonfina.

The craters of the first two are filled and destroyed by the later volcanoes, and from the Caldeira of Roccamonfina considerable craterless masses of trachytes arise. The height of the Pics is according to DEVILLE's measurement = 1430t.

128. For the edge of Caldeira de Palma I chose the highest point: Pico de los Muchachos.

Of the depth, I know only that it is very large; but the value which is in the English Maps at the bottom of the crater probably refers to the central elevation there. In order not to exaggerate, I set the depth only = 830t.

In consideration of the irregularity, of the S.W. open form of caldera, I assumed the diameter quite arbitarily to be 3100t.

129. All the details of the crater in the Azores and the Canary Islands are taken from the English sea-charts (London: published by Act of Parliament at the Hydrographic Office of the Admiralty).

These assume 1000 Fathoms = 1 geogr. Sea mile,
or 10 Cables = 1 Sea mile = 1/60 degree = 1 arc minute = 951.8 Toisen.

I chose the values of the crater for the highest crater edges, and so the values for the height of the crater bottom are given.

I determined the diameter, using the scale from the chart; values, without exception, have been transformed to use Toisen: and note that 1 Toise = 1.94904 Metre = 6.3946 Engl. feet = 6.2100 rhein. feet.

The Caldeira Sete Cidadas is located on the western end of the Azores island of San Miguel; it contains two lakes, Lagoa grande [*Lagoa Verde]* and Lagoa Azul; for the former, the numerous soundings give the greatest depth scarcely fourteen toisen.

This depth was subtracted from the sea level of the Lagoa, and compared with the Pico della Croce in the east wall of the crater, I took the depth of the caldera.

The mean sea level of the wall is probably below 300t.

In the big caldera there are four smaller, very regular craters. The wall, which is only widening to the west, seems otherwise to have a height of 330t.

The depth of the Caldeira is taken from the sea-level of the largest of the three Lagoas, which, however, have not been explored with regard to their depth. So here you have only the depth of the Lagoa Lake below the eastern wall-Pic.

131. An elliptical caldera, which contains seven hills and a Lagoa. The depth is valid for the total difference between the rather uniformly high wall and the mean sea level of the crater.

132. Monte Brazil, forming a small peninsula on the south coast of Terceira (Azores). The depth of the crater, which is not measured, and on whose base there are two round ponds, is probably very insignificant, and less than 80 toisen.

133. Two large Caldera, one in the middle, the other east on the island of Terceira, I have not considered because of their irregular shape and lack of adequate measurements. The Caldeira de Santa Barbara is located on the western edge; for the height of the wall, I took the height of the southern Pics; probably the mean height of the crater's edge is barely 500t; it encloses a very uneven depth, for which the measurements are missing.

134. The regularly circular wall surrounds an uneven basin, on the bottom of which is a Lagoa. Their height above sea level is unknown. Only the southern Pico Gorda gave a maximum of the altitude for a point of the wall. The mean height of the crater rim is perhaps hardly 400t.

135. NAUMANNs "*Lehrbuch der Geognosie*" Vol. I pg.86

136. The heights for the Tangkuban Perahu (Prahoe) are probably taken from the work of JUNGHUHNS, as well as from the Mount Merapi. The Javanese Merapi is not to be confused with the 2033t high Berapi on Sumatra.

137. Mauna Loa on Hawaii.

NAUMANN's *Geogn*. Vol. I. p. 86, the diameter = 2/5 miles, the depth = 200t (according to DOUGLAS). STRZELECKI has calculated the depth = 211t.

On the expedition of Admiral KRUSENSTERN, HORNER measured the width of the summit = 1900t from the sea. However, I only calculate 1800t for the diameter of the crater.

138. Lake Laacher at Andernach has been measured very accurately by OEYNHAUSEN. What I gave as the diameter of the crater is only the approximate average width of the former (1844) irregularly elliptically defined water surface, the largest depth of which is about 30 toisen.

The highest adjacent point of the wall over the lake is about 100t, so I guess the depth = 130t, from the summit to the bottom of the lake. As far as I know this region, after several visits, it is probable that, with the exception of one stretch, most of the valleys are about 60 to 70 toisen. The neighboring waterless mighty crater is more regularly surrounded by a rampart than Lake Laacher, but I cannot give any measurements for it.

139. The first number can be very uncertain. I have thus estimated them from Bonn against the height of the Drachenfels; also the other two values are incidental estimates that I have attempted with frequent visits of this crater.

140. NAUMANN's *Geogn*. Vol. I. p. 86, according to HOFFMANN.

141. NAUMANN's *Geogn*. Vol. I. p. 86, according to HOFFMANN.

142. Information on the depth of Vesuvius is of little value unless the time is given; the depth is very variable.

143. Pico Alto on the Azores island of Pico; it is to be understood as the chief crater in which an eruption cone with a much smaller crater rises; the data are taken from the Sea Charts from the Hydrographie Office, according to the measurements of VIDAL.

144. NAUMANN's *Geogn*. Vol. I. p. 86.

However, I choose the absolute height of the Pics of Tenerife, which was already discussed in the text, = 1910 toisen.

145. The dimension and the height of the Chahorra (Narices del Teide) are taken from the English sea-map. The depth on average according to L. v. BUCH AND DEVILLE.

146. The sea-level of Mount Etna is a fixed measurement, the height of the crater is certainly very variable.

NAUMANN does not give the depth (*Geogn*., I, p. 86).

After LEONHARD's lectures I set these = 325 feet = 54 toises (for 1830).

147. NAUMANN's Geogn. Vol. I. p. 86, according to BURKART's measurement

148. Ditto. LEONARD gives the elevation = 2443 toisen.

149. NAUMANN's (*Geogn*. Vol.I., p. 86) gives for the Rucu-Pichincha a value which belongs to the Cotopaxi or the Tolinia. Near this same value (17644 Par. Feet = 2941 toisen) has ABICH in his work: Geological observations on the volcanic phenomena, etc. p. 57. The correct height is, according to A. v. HUMBOLDT (*Kleiner Schriften* Vol. I. and Atlas No. 1) = 2490 t. The measurements of Humboldt and Von WISSE, which the latter has spent days in the pharynx of Rucu-Pichincha, can be studied in detail in order to examine the two craters of unequal ages. I estimate the depth from the highest edge to the base of the western crater.

150. NAUMANN's Geogn. Vol. I. p. 86, according to Von. GEROLT's measurement. Comp. Atlas of HUMBOLDT's, *Kl. Schriften* Tab. 8. I would have liked to add Cotopaxi and Ciflaltepetl (Pic d'Orizaba) to the table, had the dimensions of their craters been known to me.

151. It would be difficult to give an approximate list of all scriptures dealing with the dwellers of the heavenly bodies. That is of little use; most of it may be

unread because many wrote about these doubtful things lacking all the necessary knowledge. Even the elders turned their attention to these supernatural speculations.

Among the new ones, to mention only famous names, mention is made of KEPLER, HUYGENS and FONTENELLE. Thus FONTENELLE wrote dialogues on the majority of the worlds, and the learned Jesuit ATHANASIUS KIRCHER (who died in 1680) wrote his *Iter ecstaticum*, in which an unbridled imagination draws as boldly as absurd results.

Even the great HUYGENS did not consider it dignity, in his *Cosmotheoros*, which he assigned to his brother, to subject the question of the habitability of the celestial bodies to a very extensive investigation.

This work, which has already become rarer in the Latin and German texts (first printed in 1698, 3 years after HUYGENS death in the Hague), still deserves to be re-read from time to time, not to learn something new from it, but to the ideas of a profound Philosopher, a mathematical mind, to follow in this direction a little closer.

HUYGENS *Cosmotheoros* was translated into German by an unknown in 1767; the work was published (by OBELLI and GESSNER in Zurich) under the title:

"*Mr. CHRISTIAN HUYGENS's World Expert, or reasonable speculation*" etc.etc.

152. Comp. also MÄDLER, popular astronomy (ed. 1841) p. 124.

In an essay by Prorectors FISCHER:

Something of transcendent astronomy (BODE's *Jahrbuch* for 1792 p.222) has developed many of these ideas; by name, a theorem of GRAVESANDE's is illuminated, according to which the probability of the existence of inhabitants of the planets is regarded as a certainty, like one to the number of all other end purposes, to which God has been able to determine the planets.

FISCHER wrote:

"It is a very correct rule of logic that, if I recognize several cases as equally possible, the probability of the reality of any particular case will be to the certainty of how one would behave to the number of all possible cases. This rule presupposes, however, that I must know all possible cases, and this is clearly not the case in the judgment to which it applies the rule. The only possible final purpose of habitability, which I know, is the existence of inhabitants; of other possibilities, I have no concept at all, do not know at all, and can not know whether this is the only possibility, or is it still one of them. or two, or infinitely many. GRAVESANDE, therefore, had at most the absolute absolute uncertainty, complete hovering between probability and improbability, but not an infinitely small probability, that is to say, the greatest improbability. Moreover, GRAVESANDE seems to me to have gone mad with the point of view from which the whole matter must be examined by the inhabitants of other worlds. He says that the existence of real inhabitants on other worlds is inferred from their habitability; but their habitability from their likeness to the Earth."

— FISCHER finally reaches the same result, which we have touched upon in the text [p.228] that is the real strength of the conclusion that all worlds are inhabited bodies, is based on a thought that is developed or undeveloped in every human soul, and is tacitly used as an axiom with a thousand conclusions and actions, on the proposition, that organization, life, sensation, enjoyment, mental perfection, until the whole series of stages may be called, which we compress in the sole expression "animated nature," the purpose of all existence, and that lifeless nature is limited only to that there is a lively will.

— How much FISCHER is right to hope to see these and related propositions proved by the speculations which KANT has made concerning the nature of our knowledge, I cannot say.

153. HUYGENS did not yet know about the polar speckles of Mars. (See *Weltbesucher*, p. 140.)

154. According to MÄDLER (*popular. astron.),* When the flattening of the Jupiter sphere is set to 1/14, the gravity at the poles is 2.822 times, and at the equator is 2.177 times greater than that at the surface of the Earth.

155. Comp. MÄDLER, *pop. Astron.* p. 208 et seq

156. It is not to be taken too seriously, if I think here in a sentence of the fossil dinosaurs and the insects, which nevertheless seem to belong to very different geological epochs, some cases excepted, for example. of the giant salamander and of the numerous prehistoric insects of Oeningen, of which HEER has so carefully examined a large part.

157. For those of my friends, who remember this essay from Bonn, I note that it was first written in 1843, reedited in 1850, and edited into this form in February, 1855.

—To understand this description properly, but not to misunderstand or exaggerate some of its paradoxical passages, the knowledge of all the essential objects discussed in the texts cannot be dispensed with.

In selenography we call the lunar edge the eastern one, which, while the Moon is in the south (culminated), is directed towards the east or is to the left.

In our description, on the other hand, we let the Sun travel east to west, as on the Earth. (Comp. which has been mentioned in Sect./Chapter 7, where we speak of the Co-ordinate System of the Moon-sphere.)

158. The area calculated; but which is the relation between the Earth and the Moon by which the ability to reflect light is unknown.

159. Seen from the Moon, only the point on the oceanic surface of the Earth will appear very shiny, in which the image of the Sun is mirrored.

In other respects, the seas are seen to be dark, so long as they are not covered with ice in their polar parts.

— When viewed from the surface of the Earth, we usually see only the shadowed lower side of the clouds; the great splendor, which sometimes shows the edges of the clouds illuminated by the Sun, is known to everyone, and one knows that, seen from the peaks of high mountains, the clouds below are shining in the strongest almost dazzling light.

160. The Earth will appear very bright on the Moon, when the three continents of Europe, Asia, and Africa are visible; but markedly diminished when the surfaces of the Atlantic and the large oceans occupy the greater part of the disc.

161. Namely, the constellations of the fixed stars. The mutual position of two planets is also called a constellation.

162. The revolving axis of the Earth is inclined about 66½° against the elliptical plane, but the Moon is 88½°. The position of the poles of the Earth and the Moon is thus different by 22°.

163. Because the Earth revolves around its axis in 24 hours, the Moon in 709 hours turns around its axis, that is to say, in relation to the Sun.

164. The analogue of the *Lumen Secundarium*, which we observe on the night side of the Moon.

165. The far side of the Moon, where the Earth is not seen, apart from the starlight, has very dark nights.

166. This is not to be wondered at, since it is well known that even with our turbulent Northern atmosphere one is permitted to perceive our own shadows and those of houses and trees when Venus shines at her brightest.

167. I understand the meteoric phenomena so generally, after the observations of the telescopic meteors, on which I have reported on other occasions, have made the great distribution of these bodies probable to me.

168. This addition, "as far as is recognised by its ravines", cannot be avoided; From a central mountain the usual horizon can never be seen, for the summit itself lies below the level of the Moon, and a high wall of mountains surrounds the depths. The amendment will be repeated later.

169. This very ordinary phenomenon can be seen in any observation using a telescope, only with the difference that the astronomer from the distance of 50000 miles only perceives the illuminated peak of the mountain when they have been illuminated for ¼ or ½ hour. In our description, it is presupposed that these peaks are about 3-4 miles distant, and that their glowing spots are noticed in the distance from the dark sky, even though they have only a few square inches of surface.

170. I write here quite according to the impression afforded me by the total solar eclipse observed on 28 July 1851 at Rastenburg.

171. We must remember that the whole area was clearly recognizable when illuminated at night by the full Earth. That the sky cannot appear blue is already mentioned in the text.

172. All derived from our usual Moon observations at the telescope.

173. This assumption is very arbitrary, as is all that relates to the visibility of the stars during the lunar day. That you see very many stars there is arguably certain; but one does not know the limits of seeing.

174. Because the air is absent on the Moon, there can be no sound.

175. This assumption is, at any rate, quite arbitrary, but not unjustified, like a similar one of the kind somewhat earlier.
There must be no weight on it; I write as I currently believe.

176. This cresent does not span 180°, but less so from the periphery, the narrower it gets, because the diameter of the Earth appears much larger than that of the Sun.

177. That can also be doubted; perhaps the mountains, as they are covered by the shadow of the Earth, become more clearly visible in gray-red outlines the nearer they are.

178. On the Moon you cannot see the stars sparkle.

The flickering (scintillating) of the stars has its reason in the incessantly changed refraction in the atmosphere. Part of these phenomena is related to the interference of the light.

179. The light halo around the Earth will, for the most part, originate from the Earth's atmosphere, and for this very reason it will appear red: not white, like the corona lying around the Moon, during a total darkness for the Earth.

The red light of that halo is certainly the only source of the animated redness, in which we see the totally eclipsed Moon.

The comparison with a snowy landscape illuminated by the Northern lights is not dissimilar.

In December, 1840, I once saw in the surrounding area of my hometown Eutin in Holstein, quite carmin-red illuminated, when a great northern light spread so brightly over the snowy surfaces that even the Moonlight seemed to be ineffective.

180. We can not boast of our total solar eclipses; one sees only some of the larger planets and fixed stars. When I saw (28 July 1851) the last point of the Sun's rays, I saw a few seconds after the beginning of the totality, the planets Mercury and Venus brightly illuminated on either side of the Moon, surrounded by the corona.

181. This remark is based on the ever-present occurrence of the blue light at the end of total lunar eclipses.

182. An observer on the Moon will observe the nimbus about the Earth, perhaps an hour or two before and after the middle of the darkness; probably in any partial darkness, and perhaps in any conjunction, when the Earth is a few degrees north or south of the Sun.

Description of Image I (Title page)

Representation of a crater-like Moon-landscape around Tycho in oblique illumination, just before the Sun goes down.

1. the walled area Clavius.
2. the walled area Maginus.
3. the crater Tycho.
4. the walled area Longomontanus.

a b the direction of the terminator

This shadow landscape was drawn on the Sept. 17, 1843, at the Hamburger Observatory; in the position given here it corresponds to the sight using a reversing astronomical eyepiece.
It was only intended to faithfully represent the shadows cast; the brevity of time did not allow even the smaller part of the subordinate mountain members to be marked.

Description of Image I (page 24.)
Presentation of the Eastern mountains of the Mare Serenitatis, or of the Caucasus mountains with the Sun setting.

A. The Mare Serenitatis.
B. the Southern part of the Caucasus.
C. the Northern part of the Caucasus.
α. The highest Caucasus-mountain, Calippus α; height = 2980 Toisen.
D. Crater Theaetetus.
E. the ring mountain Cassini.
a b the direction of the terminator.

From an observation at the Hamburger Observatory on 17 August 1843 on a larger astronomical telescope, this reversed the images. Because of the brevity of time it was not possible to list all the small mountains; it was only intended to portray the shadow cast of Caucasus as faithfully as possible.

Printed by C.E. Elbert of Leipzig

J. F. J. Schmidt, *The Moon*, Historical & Cultural Astronomy,
https://doi.org/10.1007/978-3-030-37269-9

The manufacturer's authorised representative in the EU is Springer Nature Customer Service Centre GmbH, Europaplatz 3, 69115 Heidelberg, Germany. If you have any concerns regarding our products, please contact ProductSafety@springernature.com

Printed and bound by CPI Group (UK) Ltd, Croydon, CR0 4YY
30/09/2026
02225693-0002